Occupational Safety and Health

Fundamental Principles and Philosophies

Occupational Safety and Health

Fundamental Principles and Philosophies

Charles D. Reese

CRC Press
Taylor & Francis Group
Boca Raton London New York

CRC Press is an imprint of the
Taylor & Francis Group, an **informa** business

CRC Press
Taylor & Francis Group
6000 Broken Sound Parkway NW, Suite 300
Boca Raton, FL 33487-2742

International Standard Book Number-13: 978-1-138-74883-5 (Paperback)
International Standard Book Number-13: 978-1-138-03505-8 (Hardback)

Library of Congress Cataloging-in-Publication Data

Names: Reese, Charles D., author.
Title: Occupational safety and health : fundamental principles and philosophies / Charles D. Reese.
Description: Boca Raton : Taylor & Francis, a CRC title, part of the Taylor & Francis imprint, a member of the Taylor & Francis Group, the academic division of T&F Informa, plc, 2017. | Includes bibliographical references.
Identifiers: LCCN 2016052420 | ISBN 9781138035058 (hardback : acid-free paper) | ISBN 9781315269603 (ebook)
Subjects: LCSH: Industrial safety. | Industrial hygiene.
Classification: LCC T55 .R434 2017 | DDC 363.11--dc23
LC record available at https://lccn.loc.gov/2016052420

Visit the Taylor & Francis Web site at
http://www.taylorandfrancis.com

and the CRC Press Web site at
http://www.crcpress.com

Contents

Preface..xvii
Author.. xix

Section A   Occupational Safety and Health

1. **Introduction**..3
   Why Is Occupational Safety and Health Needed?..3
   The Components of Safety and Health Initiatives...5
   Summary..7
   Further Readings 8

2. **History** 9
   Evolution of OSH 9
   Results from History 11
   Further Readings 11

3. **Hazards**...13
   Energy...14
   Further Readings 16

4. **Occupational Safety**..17
   Summary...19
   Further Readings 20

5. **Occupational Health** 21
   Health Hazards..22
   Health Hazard Prevention...23
   Identifying Health Hazards 23
   Quick Health Hazard Identification Checklist...25
   Summary...26
   Further Readings 26

Section B   Organizing Safety and Health

6. **Manage Occupational Safety and Health**..29
   Why Management?...30
   Safety and Health (Managing)...30
   Why Is Managing Safety and Health a Needed Entity? 32
   Summary: Why Management? 34
   Further Readings 35

7. Safety and Health Programs 37
Why Have a Comprehensive Safety and Health Program?............................38
Why Build an Occupational Safety and Health Program? 39
Components of a Safety and Health Program...40
Evaluative Questions Regarding Safety and Health Programs.....................41
Tools for a Safety and Health Program Assessment 43
Why Other Required Written Programs? 45
Summary...46
Further Readings 46

8. Special Emphasis Programs 49
Summary...49
Further Readings 50

9. Accident Investigations 51
Reporting Accidents..53
Summary...54
Further Readings 55

10. Training 57
When to Train...57
Why Train New Employees? 58
Why Train Supervisors?..59
Why Train Employees?..60
Why Communications?..61
Why Is Training a Key Element? 61
Why OSHA Training? 62
Why a Legal Basis for Training?...63
Summary...63
Further Readings 64

Section C   Administration

11. Safety and Health Budget..67
Health Budgeting...68
Safety Budgeting..69
Management Budgeting...69
Environmental Budgeting 70
Product Safety Budgeting..70
Compliance Factors 70
Written Budget 71
Controlling Cost...71
Summary...72
Further Readings 72

12. Statistics and Tracking 73
Analyzing Incident Data...73
OSHA Record Keeping 74
Company Records...74
Important Ancillary Data Needed for More Complete Analysis 75
Safety and Health Statistics Data 76
Statistical Analysis for Comparisons..76
Workers' Compensation...77
Summary..77
Further Readings 78

13. Safety and Health Ethics...79
Ethics...79
Occupational Safety and Health Ethics 81
Further Readings 81

14. Employee Involvement 83
Why Should Employees Be Involved? 84
Why Does Management Have to Be Involved? 84
Why Employee Involvement?...85
Why Are Employee Outcomes Important? 86
Why Are Goals Needed for Employee Involvement?...................................87
Summary..87
Further Readings 88

15. Joint Labor/Management Safety and Health Committees............................89
Committee Makeup...90
Record Keeping...90
Do's and *Don'ts* of L/M Committees 90
Do's 90
Don'ts 91
Expectations...91
Outcomes 92
Joint L/M Occupational Safety and Health Committees..............................92
Summary..93
Further Readings 94

16. Workplace Inspections 95
Need for an Inspection...95
When to Inspect?...97
What to Inspect?..97
Types of Inspection Instruments..98
Summary..99
Further Readings 99

Section D   Personnel Involved in Occupational Safety and Health

17. Management's Commitment and Involvement...103
   Why Roles and Responsibilities?...105
   Why Management?..105
   Further Readings 106

18. Line Supervisors...107
   Further Readings 108

19. Workers..109
   Why Workers? 109
   Further Readings 111

20. Safety Director or Manager...113
   Further Readings 114

21. Safety and Health Professional 115
   Further Readings 118

22. Industrial Hygienist...119
   Why an Industrial Hygienist?...119
   When an Industrial Hygienist Is Needed 120
   Further Readings 121

23. Safety and Health Consultant..123
   Why a Need for a Consultant? 123
   Interviewing a Consultant..124
   Why a Scope of Work?...125
   The Hiring Process 126
   Summary...127
   Further Readings 128

Section E   People-Related Issues

24. Motivating Safety and Health 131
   Setting the Stage...132
   Defining Motivation 133
   Principles of Motivation..133
   The Motivational Environment 134
   Structuring the Motivational Environment...134
   Simplify Motivation...136
   The Key Person...137
   Summary...138
   Further Readings 139

25. Behavior-Based Safety.. 141
  Behavior-Based Safety Today.. 141
  BBS Described 142
  Hindrances to Implementing BBS 143
  Summary.. 144
  Further Readings 145

26. Safety Culture 147
  Defining Safety Culture... 147
  Developing or Changing a Safety Culture.. 147
  Positive Safety Culture.. 149
  Assessing Safety Culture... 150
  Summary.. 150
  Further Readings 150

27. Communicating Safety and Health 151
  The Communicator.. 152
  Communication Tools 153
    Written Materials 153
    Bulletin Boards.. 154
    Electronic Signs 154
    Computers... 154
    Safety and Health Posters 154
    Public Address System 155
    Safety Talks 155
  Summary.. 155
  Further Readings 155

28. Bullying 157
  Data Regarding Bullying 158
  Facts about Bullying 160
  Why Bullying Occurs? 161
  Why Prevention of Bullying? 162
  Summary.. 163
  Further Readings 163

29. Safety (Toolbox) Talks... 165
  Further Readings 166

30. Incentives/Rewards.. 167
  Incentives 167
  Incentive Programs 168
  Summary.. 169
  Further Readings 170

Section F   Hazards

31. Hazard Identification 173
  Hazard Identification for Protection 174
  Further Readings 175

32. Hazard Analysis 177
  Hazards and Risk 178
  Hazard Analysis 178
  Further Readings 181

33. Root Cause Analysis 183
  Summary 184
  Further Readings 185

34. Forms of Root Cause Analysis 187
  Root (Basic) Cause Analysis 188
  Root Cause Analysis Methods 188
  System Safety Engineering 189
  Further Readings 190

35. Cost 191
  The True Bottom Line 191
  Cost of Accidents 192
  Summary 192
  Further Readings 193

Section G   Risk

36. Acceptable or Tolerable Risk 197
  Why Can Risk Be Tolerated at an Acceptable Level? 197
  Further Readings 199

37. Risk Assessment 201
  Developing a Risk Assessment 201
  Purpose of Risk Assessment 202
  Explaining the Risk Assessment 202
  Risk Evaluation 202
  Summary 203
  Further Readings 203

38. Risk Management 205
  Risk Management Components 205
  Why Manage Risk? 206
  Summary 207
  Further Readings 207

Section H   Controlling Hazards

39. Designing for Prevention 211
   Designing for Humans 211
   Designers' Responsibility 212
   Further Readings 212

40. Controls 213
   Technical Aspect of Hazard Controls 213
   Source Control 214
   Control along the Path from the Hazard to the Worker 214
      Barriers 214
      Absorption 214
      Dilution 214
   Control at the Level of the Worker 214
   Selecting Controls 215
   Risk Control 215
   Evaluating the Effectiveness of Controls 215
   Awareness Devices 216
   Work Practices 216
   Administrative Controls 216
   Work Procedures, Training, and Supervision 217
   Hazard Prevention and Controls 218
   Hazard Control Summary 218
   Summary 218
   Further Readings 219

41. Personal Protective Equipment 221
   Why a Hazard Assessment? 223
   Why Establish a PPE Program? 223
   Summary 224
   Further Readings 225

42. Occupational Safety and Health Administration (OSHA) 227
   OSHA Standards 228
   Protections under the OSHAct 228
   National Institute for Occupational Safety and Health (NIOSH) 228
   Occupational Safety and Health Review Commission (OSHRC) 229
   Employer Responsibilities under the OSHAct 229
   Workers' Rights and Responsibilities under the OSHAct 230
   Discrimination against Workers 231
   Right to Information 231
   OSHA Inspections 232
   Worker Training 232
   Occupational Injuries and Illnesses 233
   Summary 233
   Further Readings 234

Section I   Accident/Incident Prevention Techniques

43. Safe Operating Procedures..237
  Why Are SOPs Beneficial?...239
  What Are the Components Needed to Develop an SOP?.................................239
  Why SOPs Are Poorly Written...240
  Why SOPs Work...241
  Summary..242
  Further Readings 243

44. Job Safety Analysis 245
  Why JSA?...245
  Why Perform a JSA/JHA?...246
  Why Four Basic Steps in Developing a JSA/JHA? 247
  Why Select Jobs by Using Criteria? 247
  Why Identify the Hazards Associated with Each Job Step?..........................248
  Why Consider Human Problems in the JSA/JHA Process?...........................249
  Why Must Ways to Eliminate or Control Hazards Be Developed?..............249
  Why Change Job Procedures?..249
  Why Change the Frequency of Performing a Job?..250
  Why Consider PPE?...250
  Summary..251
  Further Readings 251

45. Job Safety Observation...253
  Why JSO?...253
  Why Select a Job or Task for a Planned JSO? 254
  Why Preparing for a JSO Is Necessary? 255
  Why a Checklist of Activities to Be Observed Is Needed............................256
  Why the Observation?...257
  Why Conduct a Postobservation?...258
  Why Deal with Unsafe Behaviors or Poor Performance? 258
  Summary..259
  Further Readings 260

46. Fleet Safety 261
  Why Have a Fleet Safety Program?...262
  Why Vehicle Maintenance? 262
  Why Is Driver/Operator Selection Important?...263
  Why and What Records Need to Be Maintained? 264
  Summary..264
  Further Readings 265

47. Preventive Maintenance Programs...267
  Why Specific Components Are Needed for an Effective PMP?.....................268
  Why Preventive Maintenance? 268
  Why Management's Role?...268
  Why a Formalized PMP?...269

Why the Operator Should Conduct Inspections? 269
Why Maintenance? 270
Why Management's Responsibility? 270
Summary 270
Further Readings 271

Section J   Extraneous Hazards

48. Emergency Planning 275
Why EAPs? 277
Why Are Certain Elements Required or Recommended? 278
Why Preplan? 280
Summary 280
Further Readings 281

49. Workplace Security and Violence 283
Workplace Violence Statistics 283
Risk Factors 285
Prevention Strategies and Security 285
   Environmental Design 285
   Administrative Controls 286
   Behavioral Strategies 286
Perpetrator and Victim Profile 287
Cost of Violence Is a Reason to Address It 287
Prevention Efforts 287
Why Program Development and Essential Elements? 288
Types of Workplace Violence and Events 289
   Type I Events 289
   Type II Events 289
   Type III Events 290
Why Address Workplace Violence? 290
Summary 291
Further Readings 291

50. External Force: Terrorism 293
Travel Security 294
Hardening Facilities Increases Protection 294
Potential Terrorist's Weapons 295
Protection from Chemical, Biological, or Radiological Attacks 295
Summary 296
Further Readings 296

51. Off-the-Job Safety 299
Further Readings 300

Section K   Miscellaneous Safety and Health Factors

52. Human Factors 303
   Defining Human Factors 303
   Human Factor Safety...303
   Human Characteristics 304
   Why Human Causal Factors?...304
   Benefits of Using Human Factors 304
   Summary..305
   Further Readings 305

53. Ergonomics...307
   Force...307
   Repetition..308
   Awkward Postures 308
   Static Postures 308
   Vibration..308
   Contact Stress 309
   Cold Temperatures 309
   Extent of the Problem 310
   Developing an Ergonomic Program 310
   Physical Work Activities and Conditions.. 311
   Limits of Exposure.. 311
   Ergonomic Controls.. 311
   Identifying Controls 312
   Assessing Controls 312
   Implementing Controls... 313
   Tracking Progress 313
   Proactive Ergonomics... 314
   Further Readings 315

54. Product Safety 317
   Why a Safe Design Approach Should Be Employed 317
   Principles of Safe Design 317
   Legal Obligations.. 318
   Consultation 318
   When Developing a Safe Design Action Plan 318
   What Is Consumer Product Safety Screening?.. 318
   Why Are Products Screened for Safety? 319
   How Is Safety Screening Done?... 319
   Labeling—What Does It All Mean? 319
   Summary.. 319
   Further Readings 320

Section L   Safe Facilities and Handling

55. Fire Prevention and Life Safety..323
   Why Address Fire Hazards? 323
   Causes of Fires...323
   What the Occupational Safety and Health Administration Standards Require........324
   Avoiding Fires 325
   Fire Safety and Protection 326
   Fire Prevention 326
   Managing Fire Safety 327
   The Importance of Fire Safety and Life Safety Design 328
   FPP Requirements...329
   Fire Protection Summary 330
   Further Readings 333

56. Hazardous Materials..335
   Why Hazmats Need Special Handling 336
   Transportation of Hazmats 336
      Storage and Packing 337
      Transporting 337
   Regulations and Hazmats 337
   Hazardous Waste...338
   US Hazardous Waste...339
   Final Disposal of HW 339
      Recycling 340
      Portland Cement 340
      Incineration, Destruction, and Waste-to-Energy.............................340
      HW Landfill (Sequestering, Isolation, Etc.) 340
   Summary..341
   Further Readings 341

57. Transportation..343
   Employer's Responsibility 343
   Transporting Materials 343
   Employees of Transportation Businesses 344
   Summary..344
   Further Readings 345

Section M   Meshing Safety with Management Approaches

58. Process Safety Management..349
   Process Development 349
   Process Hazard Analysis..350
   Operating Procedures 351
   Safe Work Practices...352
   Employee Participation 352

Training 352
    Initial Training 352
    Refresher Training 353
    Training Documentation 353
Contractors 353
    Application 353
    Employer Responsibilities 353
Contract Employer Responsibilities 354
Pre-Start-Up Safety Review 354
Mechanical Integrity 355
Hot Work Permit 355
Management of Change 356
Incident Investigation 356
Emergency Planning and Response 357
Compliance Audits 357
Trade Secrets 357
OSHA's Response 358
Summary 358
Further Readings 359

59. Total Quality Management 361
Development of TQM in the United States 362
Features of TQM 362
Definition 363
Principles of TQM 364
TQM Implementation Approaches 365
Safety and Health Integrated into TQM 365
The Concept of Continuous Improvement by TQM 366
Steps in Managing the Transition 366
Summary 367
Further Readings 368

60. Lean Safety 369
New Approaches: Lean Safety and Sustainability 369
Benefits of Lean Safety 369
Challenges of Lean Safety 370
Changing Culture 370
Learning and Training 371
Sustainability 371
Summary 372
Further Readings 373

Index 375

Preface

Most occupational safety and health (OSH) books tell the reader how to apply all the concepts, principles, elements, and tools of prevention and develop interventions and initiatives to mitigate occupational injuries, illnesses, and deaths.

This book is not a how-to book. It is a book that addresses the principles and philosophical basis for all the varied components and elements needed to develop and manage an effective safety and health program or initiative. It is a book designed to answer the questions often posed as to why should we do it this way. It is the *why* book.

This book was developed to provide safety professionals, students, and employers with the basic tenets for the justification and purpose of an OSH initiative and for those responsible for safety and health within the industrial and business communities.

Those involved with assuring that a safe and healthy workplace is provided to employees are always challenged as to why principle and philosophical approaches are needed in order to develop and implement a functional OSH environment. Those responsible for safety and health are always answering the following questions, Why do we need this component or these elements? Can their use be justified as to effectiveness and cost? How will it work? How sure are you of your recommendation(s)?

The intent of this book is to provide a reference, blueprint, and a helpmate for the philosophical basis and the use of the principles for inclusion in OSH and the justification for their inclusion as integral components of doing business.

Suffice it to say that the protection of an employer's workforce should be the primary goal or objective of any OSH initiative.

This book includes the philosophical aspects of organizing safety and health in a corporation, company, or organization; the impacts that people have on safety and health; the tools available for application of safety and health issues; and the potential for new or evolving approaches in addressing workplace safety and health.

A philosophical approach and use of principles expresses the reasons and logic that dictate the decision-making process regarding what methodology, elements, and applications to use that have been successfully and effectively implemented during the past 46 years since the inception of the Occupational Safety and Health Administration to facilitate the protection of the United States' workforce.

Thus, *Occupational Safety and Health: Fundamental Principles and Philosophies* addresses the issue of why we address OSH issues in a structured and organized fashion in the same manner as we would use for any other facet of a business.

Author

Dr. Charles D. Reese, for 37 years has been involved with OSH as an educator, manager, or consultant. In Dr. Reese's early beginnings in OSH, he held the position of industrial hygienist at the National Mine Health and Safety Academy. He later assumed the responsibility of manager for the nation's occupational trauma research initiative at the National Institute for Occupational Safety and Health's (NIOSH) Division of Safety Research. Dr. Reese has played an integral part in trying to ensure that workplace safety and health is provided for all those within the workplace. As the managing director for the Laborers' Health and Safety Fund of North America, his responsibilities were aimed at protecting the 650,000 members of the laborers' union in the United States and Canada.

He has developed many OSH training programs, which run the gamut from radioactive waste remediation to confined-space entry. Dr. Reese has written numerous articles, pamphlets, and books on related safety and health issues.

Dr. Reese, professor emeritus, was a member of the graduate and undergraduate faculty at the University of Connecticut, where he taught courses on Occupational Safety and Health Administration (OSHA) regulations, safety and health management, accident prevention techniques, industrial hygiene, ergonomics, and environmental trends and issues. As professor of environmental/OSH, he coordinated the bulk of the environmental, safety, and health efforts at the University of Connecticut. He is called upon to consult with industry on safety and health issues and is often asked for expert consultation in legal cases.

Dr. Reese also is the principal author of the following:

Handbook of OSHA Construction Safety and Health (Second Edition)
Material Handling Systems: Designing for Safety and Health
Annotated Dictionary of Construction Safety and Health
Occupational Health and Safety Management: A Practical Approach (Third Edition)
Office Building Safety and Health
Accident/Incident Prevention Techniques (Second Edition)

The four-volume set entitled *Handbook of Safety and Health for the Service Industry:*

Volume 1: Industrial Safety and Health for Goods and Materials Services
Volume 2: Industrial Safety and Health for Infrastructure Services
Volume 3: Industrial Safety and Health for Administrative Services
Volume 4: Industrial Safety and Health for People Oriented Services

Section A

Occupational Safety and Health

For an extended period of time, occupational safety and health (OSH) has been viewed as a separate discipline, but its credibility has been enhanced by the Occupational Safety and Health Act of 1970. The intent of this discipline is centered upon providing America's workforce a safe and healthy workplace free from hazards. This section is intent upon providing a background upon which safety and health can be folded in as an integral part of the business model.

The contents of this section are as follows:

Chapter 1—Introduction
Chapter 2—History
Chapter 3—Hazards
Chapter 4—Occupational Safety
Chapter 5—Occupational Health

1

Introduction

In order to understand why occupational safety and health (OSH) is an area of concern and relevant to the operation of a business or industry, the *why* is best explained by becoming familiar with why so much emphasis is placed on the development of OSH as a critical part of the business environment.

Why businesses address OSH is as important as how they achieve an effective safety and health effort. Every business must have a reason to address OSH as well as why each component or element of the safety and health initiative or program is undertaken.

Why Is Occupational Safety and Health Needed?

Usually, consequences of not addressing safety and health have to do with why OSH is addressed by employers. These consequences include but are not limited to the following:

- Injury or illness to members of the workforce
- Loss of profit
- Loss of credibility as a responsible company
- Liability of not addressing safety and health
- Loss of productivity
- Loss of employees due to danger, risk, injury, illness, death, or unsafe/unhealthy work environment
- Decrease of employee morale
- Damage to or loss of capital investment (e.g., equipment or facilities)
- Decrease in reputation and integrity of the company
- View that the company does not exhibit good business practices

Many why questions can be posed regarding the need for OSH, such as the following:

- Why address OSH?
- Why develop an OSH program?
- Why is involvement crucial?
- Why is behavior important?
- Why is action required?
- Why are the varied components and elements of the OSH initiative undertaken?
- Why are accidents and incidents tracked?

- Why are occupational illnesses tracked?
- Why are hazards identified?
- Why are interventions and controls utilized?
- Why is training and education a part of OSH?
- Why is management of OSH important?
- Why are the Occupational Safety and Health Administration's regulations an important component?
- Why is a safe and healthy workplace important?
- Why are safety and health incentives as important as rewards and bonuses for production?
- Why is communications important for achieving OSH?

This is the first book that sets forth the principles and philosophies for OSH. In the past, most books have always undertaken to provide the tools that would help all those practitioners of OSH to develop the needed policies, procedures, and programs that are used to develop and implement an effective safety and health initiative for any type of company or organization. These have been primarily *how-to* books. Within this book, *Occupational Safety and Health: Fundamental Principles and Philosophies*, we discuss the foundational basis that motivates employers to protect their employees from injuries and illnesses caused by exposures in the workplace.

Philosophy is a critical study of the basic principles and concepts of a particular branch of knowledge—that being OSH for this book—as an integral part of conducting business while assuring that an understanding of those principles and concepts is being applied in a manner that provides a safe and healthy workplace for the workforce. This book provides the philosophical basis for application of management principles to OSH as well as the implications of culture to OSH. A short history is provided in Chapter 2 to demonstrate why OSH has become an area of workplace interest. This may help put things in perspective related to the evolution of OSH as it is known today.

The Healthy People 2010 objectives from the US Department of Health and Human Services (DHHS) has made the facts available relevant to occupational injuries and illnesses. Every 5 seconds, a worker is injured. Every 10 seconds, a worker is temporarily or permanently disabled. Each day, an average of 137 persons die from work-related diseases, and an additional 17 die from workplace trauma injuries on the job. Each year, about 70 youths under 18 years of age die from injuries at work, and 70,000 require treatment in a hospital emergency room. In 2015, the US Bureau of Labor Statistics estimated that the American workforce suffered from 2.9 million work-related injuries and illnesses, one-half of which were disabling in nature (about 1.5 million). This equates to an average of in excess of 5,900 injuries per workday. All of these statistics equate to a cost. Professor J. Paul Leigh of the University of California at Davis has seen an escalation of cost since 1992 in the amount of $33 billion. The new estimation of the cost to businesses in the United States has been estimated to have reached over $250 billion in 2012 for a 1-year period.

No matter the numbers, it demonstrates that workplace carnage is endemic and, some might say, of epidemic proportions.

A number of data systems and estimates exist to describe the nature and magnitude of occupational injuries and illnesses, all of which have advantages as well as limitations. In 1996, information from death certificates and other administrative records indicated that at least 6,112 workers died from work-related injuries. No national occupational chronic disease or mortality reporting system currently exists in this country. Therefore, scientists and policy makers must rely on estimates of the magnitude of occupational disease generated from a number of data sources and published epidemiologic (or population-based) studies. Estimates generated from these sources generally are thought to underestimate the true extent of occupational disease, but the scientific community recognizes them as the best available information. Such compilations indicate that an estimated 50,000 to 70,000 workers die each year from work-related diseases.

Current data systems are not sufficient to monitor disparities in health-related occupational injuries and illnesses. Efforts will be made over the coming decade to improve surveillance systems and data points that may allow evaluation of safety and health disparities for work-related illnesses, injuries, and deaths. Data from the National Traumatic Occupational Fatalities Surveillance System (NTOF), based on death certificates from across the United States, demonstrate a general decrease in occupational mortality over the 15-year period from 1980 to 1994. However, the numbers and rates of fatal injuries from 1990 through 1994 remained relatively stable—at over 5,000 deaths per year and about 4.4 deaths per 100,000 workers. Motor vehicle-related fatalities at work, the leading cause of death for US workers since 1980, accounted for 23% of deaths during the 15-year period. Workplace homicides became the second leading cause of death in 1990, surpassing machine-related deaths. While the rankings of individual industry divisions have varied across the years, the largest number of deaths consistently is found in construction, transportation, public utilities, and manufacturing, while those with the highest fatality rates per 100,000 workers are mining, agriculture/forestry/fishing, and construction. Data from the Bureau of Labor Statistics (BLS), Department of Labor, indicate that for nonfatal injuries and illnesses, incidence rates have been relatively stable since 1980. The rate in 1980 was 8.7 per 100,000 workers and 8.4 per 100,000 workers in 1994. Incidence varied between a low of 7.7 per 100,000 workers (1982) and a high of 8.9 per 100,000 workers (1992) over the 14-year period of 1980 to 1994.

The toll of workplace injuries and illnesses continues to harm this country. Six million workers in the United States have workplace hazards that range from falls from elevations to exposures to lead. The hazards vary dependent upon the type of industry (e.g., manufacturing) and the types of work being performed by workers (e.g., welding). The consequences of occupational accidents and incidents have resulted in pain/suffering, equipment damage, public exposure, loss of production capacity, and liability. Needless to say, these occupation-related accidents/illnesses/incidents have a direct impact upon the profit, which is commonly called the *bottom line*. The previous figures have given no indication that the need for OSH is waning. The mission to protect the American workforce from hazards within their workplace is still as germane today as in the past.

The Components of Safety and Health Initiatives

The National Safety Council (NSC) has provided guidance in the form of elements that should be addressed in order to have a successful safety and health program based on an amalgamation of research, safety and health expertise, and experience. This text was developed without the knowledge of this information's existence. These recommended elements are included in this book as well as why each should be part of the foundation for an occupational safety and health initiative.

1. Hazard recognition, evaluation, and control
2. Workplace design and engineering
3. Safety performance management
4. Regulatory compliance management
5. Occupation health
6. Information collection
7. Employee involvement
8. Motivation, behavior, and attitudes
9. Training and orientation
10. Organizational communications
11. Management and control of external exposures
12. Environmental management
13. Workplace planning and staffing
14. Assessments, audits, and evaluations

Some of the other factors affecting OSH, which should be addressed as an employer develops and implements an OSH program, are as follows:

Management factors:

Management commitment as reflected by management involvement in aspects of the safety and health program in a formal way and employers' resources committed to employers' safety and health program

Management adherence to principles of good management in the utilization of resources (people, machinery, and materials), supervision of employees, and production planning and monitoring

Designated safety and health personnel reporting directly to top management as well as duties and responsibilities from mangers, supervisors, and employees

Motivational factors:

Humanistic approach to interacting with employees

High levels of employee/supervisor contact

Efficient production planning

Hazard control factors:

Effort to improve workplace safety and health

Continuing development of employees

Clean working environment

Regular, frequent inspections

Illness and injury investigations and record-keeping factors:

Investigation of all incidents of illness and injury as well as non-lost-time accidents

Recording of all first-aid cases

There are so many elemental parts to OSH. It can be overwhelming and much too complex. But, business or industry must take the principles and philosophies that meet their individual needs and tailor them to their safety and health initiative.

The following is a list of other areas that need consideration when developing an OSH effort:

- Goal and objectives for worker safety and health
- Visibility of top management leadership
- Employee participation
- Assignment of responsibility
- Adequate authority and resources
- Accountability of managers, supervisors, and hourly employees
- Evaluation of contractor/vendor programs
- Comprehensive surveys, change analysis, routine hazard analysis
- Regular site safety and health inspections
- Employee reports of hazards
- Accident and near-miss investigations
- Injury and illness pattern analysis
- Appropriate use of engineering controls, work practices, personal protective equipment, and administrative controls
- Facility and equipment preventive maintenance
- Establishing a medical program
- Emergency planning and preparation
- Firefighting and fire prevention
- Ensuring that all employees understand hazards
- Ensuring that supervisors understand their responsibilities
- Ensuring that managers understand their safety and health responsibilities

Summary

As the introduction suggests, it is an arduous task to attempt to explain the why and wherefore of OSH. But, being able to explain or justify why the many specific factors of safety and health are needed in order to build an effective OSH program that will protect businesses' or industries' capital investment, facilities, workforce, and profitability in the global world of today is a continuous and evolving undertaking for any reputable business. Since the questions of why this or that is needed for a safety and health program are always arising, this book provides direction and guidance for this endeavor.

Further Readings

Occupational Safety and Health Administration, Website at http://www.osha.gov, Washington, DC: OSHA, 2001.

Petersen, D. *Techniques of Safety Management: A Systems Approach (Third edition)*. Goshen, NY: Aloray Inc., 1989.

Reese, C.D. *Accident/Incident Prevention Techniques (Second Edition)*. Boca Raton, FL: CRC Press, 2012.

Reese, C.D. *Occupational Health and Safety Management (Third Edition)*. Boca Raton, FL: CRC Press, 2016.

Reese, C.D. and J.V. Eidson. *Handbook of OSHA Construction Safety & Health (Second Edition)*. Boca Raton, FL: Taylor & Francis, 2006.

United States Department of Labor, Occupational Safety and Health Administration, Office of Training and Education. *OSHA Voluntary Compliance Outreach Program: Instructors Reference Manual*. Des Plaines, IL: US Department of Labor 1993.

2

History

Why is there occupational safety and health (OSH) in our present-day day world? Exploring what has transpired in the past has set the foundation of what is the modern world of OSH and most likely the future.

From the beginning of man's walk on the earth, people have realized that along with their chosen work come dangers and risk. From the hunter or gatherer, there was the danger of the environment from the natural hazards of weather, catastrophic events, and their predators or prey.

The evolution of OSH can be seen in its history. As the world changed, so did it. From the clothing that was worn every day to protective garments, we might surmise that little has changed, from the knights of the 1500s with their suits of armor to the soldiers of today with their body armor. Of course, the warrior of today has the refinements of modern technology.

But, history has taught us that our failure to consider it will result in us repeating its lessons.

Evolution of OSH

As iterated previously, it did not take long for workers and others to realize that with their work came inherent hazards that were unique to their profession. These could lead to injury and death. Also, there were exposures to substances in their workplace that also could cause both acute and chronic illnesses as well as death. From early on, these potential hazards and dangers were identified as an occupational risk.

Historically, the Egyptians were aware of the dangers from gold and silver fumes. They even had a first-aid manual for workers as early as 3000 BC. In 2000 BC, Hammurabi placed a value on permanent injuries, such as the loss of an eye, for which the owner paid the worker or paid the doctor's bill. In 1500 BC, Ramses hired a physician for quarry workers. In 400 BC, Hippocrates, the father of medicine, realized stonecutters were having breathing problems. In 100 BC, the Romans were aware of the dangers faced by workers. They would free a slave if he/she survived the launching of a ship. The Romans even had a goddess of safety and health named *Salus*, whose image was often embedded on their coins.

In the Middle Ages, people became more aware of the link between the type of work they did and the types of injuries and illnesses that they sustained. For example, English chimney sweeps in the 1700s were more susceptible to testicular cancer because of the soot. With the advent of the industrial revolution, the use of machinery and the changed work environment saw a rapid rise in the number of injuries, illnesses, and deaths. During this period, the first unions began to be organized to try to protect workers from the hazards of the workplace. The only improvement in the 1800s was fire protection because of pressure from insurance companies. This was soon followed by Massachusetts' requirement for factory inspections. Also, the first Acts and Regulations pertaining to mining were introduced. Some safety measures were adopted for other industries such as the railroads with the invention of air brakes and automatic couplers, which saved many lives and amputations.

During the first part of the 1900s, workers' compensation laws started appearing and were finally deemed constitutional by the Supreme Court in 1916. Before this, most employers blamed their workers and held them responsible for workplace incidents, citing what were known as the "common laws," which stated the following:

1. Employer is not responsible when a fellow worker negligently causes your injury.
2. Employer is not responsible if the worker is injured due to his/her own negligence.
3. If an employee takes up a risky job knowing fully well the inherent hazards, the employer is not responsible.

Under the workers' compensation laws, the employers assumed responsibility for their workplaces' safety and health. They were required to provide and pay for medical care and lost wages due to on-the-job incidents. Also at this time, interest was generated in counting the numbers of injuries and deaths; the most famous of these undertakings was the work-related death count for Allegheny County, Pennsylvania, by the Russell Sage Foundation.

It was during this time that mining catastrophes continued to occur and more laws were passed to protect miners. Some catastrophes, such as that at the Triangle Shirtwaist Co. in 1910, where 146 young women were killed in a fire because exit doors were locked, demonstrated a need to better protect workers. When 2,000 workers or 50% of the workforce died from silica exposure at Gauley Bridge, West Virginia, the Walsh–Healey Act was passed that required safety and health measures for any employer receiving a government contract. Some companies began to understand their moral responsibility. The American Match Co. allowed other companies in the match-producing industry to use their process, which substituted a safer substance for phosphorus in match making. This resulted in the decrease of an occupational illness called *phossy jaw*, which caused swelling and pain in the jaw due to phosphorus exposure. A more detailed timeline related to OSH can be found in *Occupational Health and Safety: A Practical Approach (Third Edition).* The chronology brings the steps in the evolution of OSH up to date.

As pressure mounted from workers and unions to pass some federal laws and the number of injuries, illnesses, and deaths increased, it became more apparent that the state programs for OSH were not protecting workers effectively. If it were not for unions attempting to protect their members, the Occupational Safety and Health Act of 1970 would probably not have passed, and workers would have much less protection from job hazards today. Now, most employers have realized that a safe and healthy workplace is more productive and makes good business sense.

There are six good reasons to prevent occupational accidents, injuries, illnesses, and deaths:

1. Destruction of human life is morally unjustified.
2. Failure of employers or workers to take precautions against occupational injuries and illnesses makes them morally responsible for these incidents.
3. Occupational incidents limit efficiency and productivity.
4. Occupational accidents and illnesses produce far-reaching social harm.

5. Safety techniques have produced reduction of accident rates and severity rates.
6. Recent cries and mandates have come forth at the state and federal levels to provide a safe and healthy workplace.

Results from History

All of this history has resulted in the development of our modern-day safety and health approach with all of its various components and elements.

All of this history has driven our approach to OSH today and is most likely the reason that we experience fewer overall

- Catastrophic events
- Occupational injuries
- Occupational illnesses

But, having OSH has had significant effects on workplace safety and health. Although employment has almost doubled, workplace fatalities have gone down by more than 65% since the inception of the Occupational Safety and Health Administration (OSHA). There were about 38 worker deaths a day in 1970 as compared to 13 per day in 2011. Occupation injury and illness rates have also lowered noticeably. In 1972, the workforce experienced about 10.9 incidents per 100 workers as compared to fewer than 4 per 100 in 2010.

These results are why there is a more organized and professional approach to OSH today. This is why history has been an important precursor and driver of safety and health as an integral part of the business model for today's businesses and industries.

Safety and health has been legitimized as a critical mainstay in businesses' or industries' continuous striving to improve their bottom line to remain competitive and sustainable.

Further Readings

Reese, C.D. *Accident/Incident Prevention Techniques (Second Edition)*. Boca Raton, FL: CRC Press, 2012.
Reese, C.D. *Occupational Health and Safety Management: A Practical Approach (Third Edition)*. Boca Raton, FL: CRC Press, 2016.

3

Hazards

There would not be a need for occupational safety and health (OSH) as a discipline if hazards did not exist in the workplace. Businesses, industries, and companies are formed to provide services, goods, and manufactured items; develop new products; provide assistance; construct edifices; or transport or move materials. With these entities come process, procedures, and outcomes that include a multitude of hazards.

Each of the previous named business activities generates a myriad of hazards. Each hazard or set of hazards is unique to each business or industry environment.

Hazards are defined as a source of danger that could result in a chance event such as an accident/incident. A danger itself is a potential exposure or a liability to injury, pain, or loss. Not all hazards and dangers are the same. Exposure to hazards may be dangerous, but this is dependent upon the amount of risk that accompanies it. The risk of water contained by a dam is different compared to being caught in a small boat in rapidly flowing water. Risk is the possibility of loss or injury/illness or the degree of the possibility of such loss. Incidents do not occur if a hazard does not exist that presents a danger to those working around it. If the potential exposure is high, there is a greater the risk that an undesired event will occur.

The advent of the identification of hazards in the workplace had it origins in the early evolution of OSH. The reason that hazards became major elements regarding safety and health in the workplace was the realization that they posed a risk to those in the workplace. This risk was of two major types, either a safety risk, which can result in trauma injuries or death, or a health risk, which can result in illnesses or death from illness.

Each hazard must be addressed separately because of their unique sources and effects.

Most accidents/incidents are caused by the unplanned or unwanted release of large amounts of energy, or by the release of hazardous materials. In a breakdown of accident causes, the direct cause of these unplanned events is the energy or hazardous material released at the time of the incident. If no one can identify these sources of energy, then there is no way to control or prevent their release. Thus, the hazards will exist, and the end result will be less than favorable.

There are reasons why the sources of energy need to be identified in the work environment. Why identify these hazards?

- They often release tremendous amounts of stored energy.
- At times, they are not recognized as a hazard.
- They are a hazard based on their size and weight.
- Their motion generates large amounts of energy.
- Combining them with other sources or materials increases their hazard risk.
- They can be better controlled.
- The release of energy can have disastrous effects on equipment, workers, and the environment.

Energy

Energy is classified in one of two ways. It is either potential or kinetic energy. Potential energy is defined as stored energy such as a rock on the top of a hill. There are usually two components to potential energy. They are the weight of the object and its height above another surface. The rock resting at the bottom of the hill has little potential energy as compared to the one at the top of the hill. Some examples of potential energy are as follows:

Examples of Potential Energy	
Compressed gases	Hand or power tool
Object at rest	Liquefied gas
Effort to move an object	Dust
Spring-loaded objects	Unfallen tree
Electrically charged component	Radiation source
Idling vehicle	Chemical source
Disengaged equipment	Biological organism
Flowable material	

The other classification is kinetic energy, which is best described as energy in motion. Kinetic energy is dependent upon the mass of the object. Mass is the amount of matter making up an object; for instance, an elephant has more matter than a mouse and therefore more mass. The weight of an object is a factor of the mass of an object and the pull of gravity on it. Kinetic energy is a function of an object's mass and its speed of movement or velocity. A bullet thrown at someone has the same mass as one shot at a person, but the difference is in the velocity, and there is not any disagreement as to which has the most kinetic energy or potential to cause the greatest damage or harm. Some examples of kinetic energy are as follows:

Examples of Kinetic Energy	
Operating tools or equipment	Moving conveyors
The flow of materials	Running machines
Falling objects	Running equipment
Lifting a heavy object	Moving dust
Moving vehicles or heavy equipment	Tree falling
Release of energy from radiation, chemical, or biological sources	Pinch area from moving objects
Energy transfer devices such as pulleys, belts, gears, shears, edgers	Running power tools

Energy comes in many forms, and each has its own unique potential for release of energy and its own ensuing form of danger and potential for an accident or incident that can result from the energy being released. The forms of energy are pressure, biological, chemical, electrical, thermal, light, mechanical, and nuclear. Examples of each form of energy are as follows:

Examples of Sources of Energy						
Pressure Energy	Biological Energy	Chemical Energy	Electrical Energy	Light Energy	Nuclear Energy	Thermal (Heat) Energy
Pressurized vessel	Allergens	Corrosive materials	Capacitors	Intense light	Alpha particles	Chemical reactions
Caisson work	Biotoxins	Flammable/combustible materials	Transformers	Lasers	Beta particles	Combustible materials
Explosives	Pathogens	Toxic chemicals	Energized circuits	Infrared sources	High-energy nuclear particles	Cryogenic materials
Noise	Poisonous plants	Compressed gases	Power lines	Microwaves	Neutrons	Fire
Compressed gases		Carcinogens	Batteries	Sun	Gamma rays	Flames
Steam source		Confined spaces	Exposed conductors	Ultraviolet light	X-rays	Flammable materials
Liquefied gases		Oxidizing materials	Static electricity	Welding		Friction
Air under pressure		Reactive materials	Lightning	Radio Frequency fields		Hot processes
Diving		Poisonous chemicals and gases		Radio frequency		Hot surfaces
Confined spaces		Explosives				Molten metals
		Acids and bases				Steam
		Oxygen deficiency atmosphere				Solar
		Fuels				Weather phenomena
		Dusts or powders				Welding

If the direct cause type of energy is known, then equipment, materials, and facilities can be redesigned to make them safer, personal protection can be provided to reduce injuries, and workers can be trained to be aware of hazardous situations so that they can protect themselves against them.

Each of these poses types of energy its own unique danger and also varies greatly in the degree of risk posed as well as the type of energy that can be released when not prevented or controlled. All this energy with its potential to be released in the workplace can result in a variety of incidents or accidents. This is another reason why hazards are a detriment to workers in the workplace.

Further Readings

Reese, C.D. *Accident/Incident Prevention Technique (Second Edition)*. Boca Raton, FL: CRC Press/Taylor & Francis Group, 2011.

Reese, C.D. *Occupational Health and Safety Management: A Practical Approach (Third Edition)*. Boca Raton, FL: CRC Press/Taylor & Francis Group, 2016.

Reese, C.D. and J.V. Eidson. *Handbook of OSHA Construction Safety & Health (Second Edition)*. Boca Raton, FL: CRC/Taylor & Francis Group, 2006.

Saskatchewan Labour, *Identifying and Assessing Safety Hazards*. http://www.labour.gov.sk.ca/safety, 2007.

United States Department of Labor. Occupational Safety and Health Administration. Office of Training and Education. *OSHA Voluntary Compliance Outreach Program: Instructors Reference Manual*. Des Plaines, IL: US Department of Labor, 1993.

United States Department of Labor. Occupational Safety and Health Administration. *Field Inspection Reference Manual (FIRM): OSHA Instruction CPL 2.103*. Washington, DC. September 26, 1994.

United States Department of Labor, Mine Health and Safety Administration. *Hazard Recognition and Avoidance: Training Manual*. MSHA 0105, Revised May 1996.

4

Occupational Safety

The errant or unplanned release of energy results in workers being injured or even killed.

There are such a myriad of safety hazards facing employers and workers within the workplace. It is difficult to select the most important hazards and thus err by leaving out ones that others believe to be important.

Trauma is, by definition, an injury produced by a force (violence, thermal, chemical, or an extrinsic agent). Occupational trauma transpires from the contact with or the unplanned release of varied sources of energy intrinsic within the workplace. Most workplaces are a plethora of energy sources from potential (stored) energy to kinetic (energy in motion) energy sources. These sources may be stacked materials (potential) or a jackhammer (kinetic). It is the sources of energy that are the primary causes of trauma deaths and injuries to workers.

Trauma-related events are a lot easier to observe and evaluate. This is why trauma events are often given the most emphasis.

- Trauma occurs in real time with no latency period.
- Trauma cases have an immediate sequence of events.
- Outcomes are readily observable (only a few minutes or hours have to be reconstructed).
- Root or basic causes are more clearly identified.
- Events allow for easy detection of cause-and-effect relationships.
- Deaths and injuries are not difficult to diagnose.
- Deaths and injuries are highly preventable.

When the Bureau of Labor Statistics released the results of the annual workplace injury and illness summary for 2010, it indicated that traumatic workplace injuries numbered 2.9 million (94.9%) of the total workplace injuries and illnesses (3.1 million). The National Census of Fatal Occupational Injuries for 2010 indicated that 4,547 workers died in the workplace. These numbers are provided to indicate why occupational safety deserves attention within the American workplace.

Why does occupational safety as a term merit a focus point in occupational safety and health? Because it is the following:

- Doing things in a manner so that no one will get hurt and so the equipment and product will not get damaged
- The implementation of good engineering design, personnel training, and the good common sense to avoid bodily or material harm
- The systematic planning and execution of all tasks so as to produce safe products and services with relative safety to people and property

- The protection of persons and/or equipment from hazards which exceed normal risk
- The application of techniques and controls so as to minimize the hazards involved to a particular event or operation—with consideration to potential personal injury and/or damage
- A process employed to prevent accidents both in the area of conditioning the environment as well as conditioning the person toward a safe behavior
- The way to function with minimum risk to personal well-being and to property
- The art of controlling exposure and/or hazards that could cause personal injury and property damage
- Controlling people, machines, and/or the environment that could cause personal injury or property damage
- Performing daily tasks at the workplace in the safe manner in which they should be done, or when this is not possible or a lack of knowledge exists, seeking the necessary knowledge and assistance
- The elimination of foreseen hazards and the necessary training to prevent accidents or provide limited acceptable risk to personnel and facilities

The reasons why occupational safety is needed can be viewed and described in many different ways. Nearly all of these definitions include property damage as well as personal injury. It shows that the thinking is in the right direction—that safety consists of a total loss-control activity.

Since the causes of accidents/incidents are known, there's no such thing as new accidents, just new victims.

There are two forms of indirect cause. They are unsafe conditions and unsafe acts. Unsafe acts are relevant to workers' behavior, which will be addressed in later parts of this book. There are many factors that have the potential to be safety hazards. See the following list for some examples of safety hazards.

Sources of Safety Hazards	
Abrasives	Emergencies
Acids	Environmental factors
Biohazards	Excavations
Blasting	Explosives
Blood-borne pathogens	Falls
Caustics	Fibers
Chains	Fires
Chemicals	Flammables
Compressed-gas cylinders	Forklifts
Confined spaces	Fumes
Conveyors	Gases
Cranes	Generators
Derricks	Hand tools
Electrical equipment	Hazardous chemical processes
Elevators and manlifts	Hazardous waste

(Continued)

Sources of Safety Hazards	
Heavy equipment	Pressure vessels
High voltage	Radiation
Hoists	Respirators
Hoses	Rigging
Hot items	Scaffolds
Hot processes	Slings
Housekeeping/waste	Solvents
Ladders	Stairways
Lasers	Storage facilities
Lifting	Stored materials
Lighting	Transportation equipment
Loads	Transportation vehicles
Machines	Trucks
Materials	Ventilation
Mists	Walkways and roadways
Noise	Walls and floor openings
Platforms	Welding and cutting
Power sources	Wire ropes
Power tools	Working surfaces

Summary

Some basic principles are listed in the table below.

Five Principles of Safety
1. All accidents are preventable.
2. All levels of management are responsible for safety.
3. All employees have the responsibility to themselves, their coworkers, and their family to work safely.
4. In order to eliminate accidents, management must ensure that all employees are properly trained on how to perform every job task safely and efficiently. Knowledge is the key to safety.
5. Every employee must be involved in every area of the safety and production process. People like to be involved in the decisions that affect them.

Safety awareness is not automatically attributed to the work force. It must be carefully developed if we truly care. Safety doesn't happen by accident. Accidents hurt—safety doesn't.

Further Readings

Reese, C.D. *Accident/Incident Prevention Techniques (Second Edition)*. Boca Raton, FL: CRC Press/Taylor & Francis Group, 2011.

Reese, C.D. *Occupational Health and Safety Management: A Practical Approach (Third Edition)*. Boca Raton, FL: CRC Press/Taylor & Francis Group, 2016.

Reese, C.D. and J.V. Eidson. *Handbook of OSHA Construction Safety & Health (Second Edition)*. Boca Raton, FL: CRC/Taylor & Francis Group, 2006.

Saskatchewan Labour. *Identifying and Assessing Safety Hazards*. http://www.labour.gov.sk.ca/safety, 2007.

United States Department of Labor. Occupational Safety and Health Administration. Office of Training and Education. *OSHA Voluntary Compliance Outreach Program: Instructors Reference Manual*. Des Plaines, IL: US Department of Labor, 1993.

United States Department of Labor. Occupational Safety and Health Administration. *Field Inspection Reference Manual (FIRM): OSHA Instruction CPL 2.103*. Washington, DC. September 26, 1994.

5

Occupational Health

The 372,000 occupational illnesses include repeat trauma such as carpal tunnel syndrome, noise-induced hearing loss, and poisonings. It is suspected that many occupational illnesses go unreported when an employer or worker is not able to link exposure with the symptoms the employees are exhibiting. Also, physicians fail to ask the right questions regarding the patient's employment history, which can lead to the commonest of diagnoses, a cold or flu, rather than an occupationally related illness or exposure. This has become very apparent with the recent occupational exposure to anthrax where a physician sent a worker home with anthrax without addressing his/her potential occupational exposure hazards. Unless physicians are trained in occupational medicine, they seldom address work as the potential exposure source.

This is not entirely a physician problem by any means, since the symptoms that are seen by the physician are often those of flu and other common illnesses suffered by the general public. It is often up to employees to make the physician aware of their on-the-job exposure. If you notice, the term that has been continuously used is *exposure* since, unlike trauma injuries and deaths, which are usually caused by the release of some source of energy, occupational illnesses are often due to both short-term and long-term exposures. If the results of an exposure lead to immediate symptoms, it is said to be acute. If the symptoms come at a later time, it is termed a chronic exposure. The time between exposure and the onset of symptoms is called the latency period. It could be days, weeks, months, or even years, as in the case of asbestos, where asbestosis or lung cancer appears 20 to 30 years after exposure. After looking at large numbers of death certificates (20,000) from specific groups of workers, there is often seen a significant number of cases of specific types of cancer such as liver, thyroid, or pancreatic cancer that do not appear in the same number in the normal adult population. This leads one to believe that something the workers were exposed to in the work environment may have caused their demise.

It is often very difficult to get employers, supervisors, and employees to take seriously the exposures in the workplace as a potential risk to the workforce both short and long term, especially long term. "It can't be too bad if I feel all right now." This false sense of security is illustrated by the 90,000 occupational illness deaths that are estimated by the Bureau of Labor Statistics to occur each year. This far surpasses the 6,000 occupation trauma deaths a year. If both trauma and illness deaths are added together, they would be equivalent to the lives lost to a jumbo jet crashing every day of the year. Would an aviation record like this be acceptable? I doubt that many of us would be flying. It is time for employers and the workforce to take on-the-job exposures as a potentially serious threat.

It is widely believed that many of the ill effects that workers suffer are, in all likelihood, linked to their chemical exposure at work. But, not as much has been made of the chemical exposure to office workers in their containerized environment, where chemical vapors, mist, and particles can be spread throughout the building by the ventilation system. Only in recent years has the term *sick building syndrome* become a commonly used term.

Health Hazards

Health hazards are caused by any chemical or biological exposure that interacts adversely with organs within our body, causing illnesses or injuries. The majority of chemical exposures result from inhaling chemical contaminants in the form of vapors, gases, dusts, fumes, and mists, or from skin absorption of these materials. The degree of the hazard depends on the length of exposure time and the amount or quantity of the chemical agent. This is considered to be the dose of a substance. A chemical is considered a poison when it causes harmful effects or interferes with biological reactions in the body. Only those chemicals that are associated with a great risk of harmful effects are designated as poisons.

Dose is the most important factor determining whether or not someone will have an adverse effect from a chemical exposure. The longer someone works at a job and the more chemical agent that gets into the air or on the skin, the higher the dose potential. Two components that make up dose are as follows:

- The length of exposure, or how long someone is exposed—1 hour, 1 day, 1 year, 10 years, etc.
- The quantity of substance in the air (concentration), how much someone gets on the skin or breathes into the lungs, and/or the amount eaten or ingested.

Another important factor to consider about the dose is the relationship of two or more chemicals acting together that cause an increased risk to the body. This interaction of chemicals that multiply the chance of harmful effects is called a synergistic effect. Many chemicals can interact, and although the dose of any one chemical may be too low to affect someone, the combination of doses from different chemicals may be harmful. For example, the combination of chemical exposures and a personal habit such as cigarette smoking may be more harmful than just an exposure to one chemical. The combination of smoking and exposure to asbestos increases the chance of lung cancer by as much as 50 times.

The type and severity of the body's response is related to the dose and the nature of the specific contaminant present. Air that looks dirty or has an offensive odor may, in fact, pose no threat whatsoever to the tissues of the respiratory system. In contrast, some gases that are odorless, or at least not offensive, can cause severe tissue damage. Particles that normally cause lung damage can't even be seen. Many times, however, large visible clouds of dust are a good indicator that smaller particles may also be present.

The body is a complicated collection of cells, tissues, and organs having special ways of protecting itself against harm. These are usually called the body's defense systems. The body's defense systems can be broken down, overcome, or bypassed. When this happens, injury or illness can result. Sometimes, job-related injuries or illness are temporary, and someone can recover completely. Other times, as in the case of chronic lung diseases like silicosis or cancer, these are permanent changes that may lead to death.

Health Hazard Prevention

Occupational illnesses have some unique features that can be attributed to the under-reporting of them. The following is some information about occupational illnesses:

- They are more difficult to recognize or diagnose.
- There is often a latency period between exposure and the occurrence of symptoms.
- There is not always a clear cause-and-effect relationship.
- Root or basic causes are not clearly apparent.
- It is difficult to determine a sequence of events.

This is why preventing occupation illnesses is more difficult to address in the workplace environment. Occupational illnesses are often exacerbated by exposures off the job.

As with any problem, goals must set forth in order to address long-term prevention. In the arena of prevention, the objectives for Healthy People 2020 are as follows:

- Reduce the rate of injury and illness cases involving days away from work due to overexertion or repetitive motion.
- Reduce pneumoconiosis.
- Reduce the proportion of workers with elevated blood lead concentration from occupational exposure.
- Reduce occupation skin diseases or disorders among full-time workers.
- Reduce new cases of work-related noise-induced hearing loss.
- Increase the proportion of employees who have access to workplace programs that prevent or reduce employee stress.

Identifying Health Hazards

A health hazard is any agent, situation, or condition that can cause an occupational illness. There are five types:

- Chemical hazards, such as battery acid and solvents.
- Biological hazards, such as bacteria, viruses, dusts, and molds. Biological hazards are often called *biohazards*.
- Physical agents (energy sources) strong enough to harm the body, such as electric currents, heat, cold, light, vibration, noise, and radiation.
- Work design (ergonomic) hazards.
- Workplace stress.

Health-related hazards must be identified (recognized), evaluated, and controlled to prevent occupational illnesses that come from exposure to them. Health-related hazards come in a variety of forms, such as chemical, physical, ergonomic, or biological:

1. Chemical hazards arise from excessive airborne concentrations of mists, vapors, gases, or solids that are in the form of dusts or fumes. In addition to the hazard of inhalation, many of these materials may act as skin irritants or may be toxic by absorption through the skin. Chemicals can also be ingested, although this is not usually the principal route of entry into the body.

2. Physical hazards include excessive levels of nonionizing and ionizing radiations, noise, vibration, and extremes of temperature and pressure.

3. Ergonomic hazards include improperly designed tools or work areas. Improper lifting or reaching, poor visual conditions, or repeated motions in an awkward position can result in accidents or illnesses in the occupational environment. Designing the tools and the job to be done to fit the worker should be of prime importance. Intelligent application of engineering and biomechanical principles is required to eliminate hazards of this kind.

4. Biological hazards include insects, molds, fungi, viruses, vermin (birds, rats, mice, etc.), and bacterial contaminants (sanitation and housekeeping items such as potable water, removal of industrial waste and sewage, food handling, and personal cleanliness can contribute to the effects from biological hazards). Biological and chemical hazards can overlap.

5. Workplace stress is anything that impacts the health of workers and is part of the overall work environment, and it is considered by most professionals to be ergonomically related and to put the worker at risk of accidents and stress-related health problems. This type of stress may be from job expectations, extremes of pressure from supervisors and peer pressure, bullying and harassment, as well as shift work or excess overtime; these can seriously harm the health and well-being of workers. Stress can also interfere with efficiency and productivity.

Shift workers have irregular patterns of eating, sleeping, working, and socializing that may lead to health and social problems. Shift work can also reduce performance and attentiveness. In turn, this may increase the risk of accidents and injuries. Statistics suggest that certain shift workers (such as employees in convenience stores and other workplaces that are open 24 hours a day) are more likely to encounter violent situations when working alone.

A health hazard may produce serious and immediate (acute) effects and symptoms. It may cause long-term (chronic) problems or may have a long period between exposure and the occurrence of the disease or illness (latency period). All or part of the body may be affected. Someone with an occupational illness may not recognize the symptoms immediately. For example, noise-induced hearing loss is often difficult for victims to detect until it is advanced. This is why it is important to identify potential health hazards in the workplace.

Finding health hazards is an investigative process entailing a systematic approach that requires that many facets and information need to be reviewed, such as the following:

- Prepare a list of known health hazards in the workplace based upon records and events.
- Review the total facility, floor plans, and work process diagrams to identify health hazard sources and locations.
- Interview workers, supervisors, and managers to identify known and suspected health hazards not already on the list.
- Make use of the five senses, and use an industrial hygienist if validation of your observations are needed. The industrial hygienist can perform accurate sampling as well as give expert advice.

Quick Health Hazard Identification Checklist

This is also why each of these questions needs to be answered:

- What chemical substances are produced, used, handled, stored, or shipped in the workplace?
- Are any vapors, gases, dusts, mists, or fumes present (including chemical by-products of work processes)?
- Are biological substances (such as bacteria, viruses, parasites, dusts, molds, and fungi) present in the workplace, the ventilation systems, and other components of the physical plant?
- Are physical agents (energy sources strong enough to harm the body, such as electric currents, heat, light, vibration, noise, and radiation) present?
- Are temperature extremes present?
- Do ergonomic hazards exist—such as work requiring lifting, awkward posture, repetitive motions, excessive muscular force, or computer use?
- Could any work processes, tools, or equipment cause health hazards (such as back injuries, soft tissue injuries, whole-body vibration, hearing loss, infections, and so forth)?
- Could departures from safe work practices cause illnesses?
- Can any potential health hazards be detected with the senses (smell, taste, touch, hearing, and sight)?
- Is there a presence of harmful stress in the workplace?
- Are there any complaints from workers about workplace-related health problems?

Summary

The employer and workers can work together to identify and control health hazards by assessing the risks to workers' safety and health posed by the work; workers should be informed about the nature and extent of the risks and how to eliminate or reduce them.

Further Readings

Reese, C.D. *Material Handling Systems: Designing for Safety and Health*. New York: Taylor & Francis, 2000.

Reese, C.D. *Industrial Safety and Health for Infrastructure Services*. Boca Raton, FL: CRC Press, 2009.

Reese, C.D. *Accident/Incident Prevention Techniques (Second Edition)*. Boca Raton, FL: CRC Press, 2012.

Reese, C.D. *Occupational Health and Safety Management (Third Edition)*. Boca Raton, FL: CRC Press, 2016.

Reese, C.D. and J.V. Eidson. *Handbook of OSHA Construction Safety & Health (Second Edition)*. Boca Raton, FL: CRC/Lewis Publishers, 2006.

United States Bureau of Labor Statistics, http://www.bls.gov, Washington, DC: US BLS, 2013.

United States Department of Labor, Occupational Safety and Health Administration. *OSHA 10 and 30 Hour Construction Safety and Health Outreach Training Manual*. Washington, DC: US Department of Labor, 1991.

United States Department of Labor, Occupational Safety and Health Administration, Office of Training and Education. *OSHA Voluntary Compliance Outreach Program: Instructors Reference Manual*. Des Plaines, IL: US Department of Labor, 1993.

United States Department of Labor, Occupational Safety and Health Administration. *General Industry Digest (OSHA 2201)*. Washington, DC: GPO, 2012.

United States Department of Labor. Occupational Safety and Health Administration. Subject Index. "Internet." 2013. Available at http://www.osha.gov.

United States Department of Labor, Occupational Safety and Health Administration. 29 Code of Federal Regulations 1910. Washington, DC: GPO, 2013.

United States Department of Labor, Occupational Safety and Health Administration. General Industry. Code of Federal Regulations. Title 29, Part 1910. Washington, DC: GPO, 2013.

Section B

Organizing Safety and Health

Like anything else in business, occupational safety and health (OSH) needs direction and guidance. It also needs a foundation of policies and procedures that will all be managed in an effective manner to achieve the business' safety and health goals and objectives.

The contents of this section are as follows:

Chapter 6—Manage Occupational Safety and Health

Chapter 7—Safety and Health Programs

Chapter 8—Special Emphasis Programs

Chapter 9—Accident Investigations

Chapter 10—Training

6

Manage Occupational Safety and Health

Management is the organizing and controlling of the components of a business or a sector of a business or industry. It is the act of handling or controlling something successfully. It is also the organization and coordination of the activities of a business in order to achieve defined objectives Management in business and organizations is the function that coordinates the efforts of people to accomplish goals and objectives using available resources efficiently and effectively. Management comprises planning, organizing staffing, leading or directing, and controlling an organization or initiative to accomplish goals or objectives. Resourcing encompasses the deployment and manipulation of human resources, financial resources, technological resources, and natural resources for the benefit of the company, employees, and employer. Management is included as one of the factors of production—along with machines, materials, and money. As a discipline, management comprises the interlocking functions of formulating corporate policy and organizing, planning, controlling, and directing a firm's resources to achieve a policy's objectives. This implies effective communication: an enterprise environment (as opposed to a physical or mechanical mechanism) implies human motivation and some sort of successful progress or system outcome. As such, management is nor the manipulation of a mechanism (machine or automated program), nor the herding of animals, and can occur in a legal as well as an illegal enterprise or environment. Based on this, management must have humans, communication, and a positive enterprise endeavor. Plans, measurements, motivational psychological tools, goals, and economic measures (profit, commanding, etc.) may or may not be necessary components for there to be management. Management consists of six functions:

1. Forecasting
2. Planning
3. Organizing
4. Commanding
5. Coordinating
6. Controlling

Using this short, concise definition of management sets the foundation for the discussion as to why occupational safety and health (OSH) should be managed as a component of any organization or industry.

Why Management?

OSH is like any other company function. Thus, it needs to be managed. Managed means that it needs a direction, a purpose, resources, and commitment.

Why manage OSH? The primary reason for managing OSH is forged in the responsibility of employers to provide to their employees a workplace free from known or anticipated safety and health hazards.

What does that entail? The assumption that the ultimate responsibility for providing for the safety and health of the workforce rests in management. Commitment, budgeting, and planning are all management responsibilities.

Managing OSH relies on all levels of management, including the elicitation of help from the workforce. If the president or chief executive officer (CEO) has not committed to providing a safe and healthy workplace, then all is lost, and a true effort to manage OSH is doomed to failure. This is why top management must demonstrate a managerial commitment to OSH. The type of commitment needed is more than just a verbal announcement. It should be in writing (signed) and supported by actions such as obeying OSH rules, providing a budget for OSH, and designating an individual (e.g., safety director) to coordinate and manage the company's safety and health initiative.

The failure to effectively manage safety and health manifests itself in the form of unsafe or unhealthy work situations or conditions and unsafe and unhealthy work performance by the workforce. This is another reason why management of OSH is a necessary part of doing business. The failure to manage the safety and health function of a company most often has had a deleterious effect upon the bottom line. In simple terms, failure to manage safety and health costs dollars.

From experience, it is recognized that accidents and incidents related to OSH are the result of management's inability to manage safety and health as they would manage any other company function.

There is the question as to why both supervisor and workers need to be involved in a management approach to OSH. This type of involvement will benefit management's safety and health initiative because it fosters ownership by stimulating investment, responsibility, and accountability, all of which results in integrating both supervisors and workers into an effective safety and health management approach.

The involvement of managers and supervisors as a function of management assures the communication of safety and health policies and procedures. This should result in structuring an environment for holding these individuals accountable and providing them the authority to implement and enforce the intent of upper management.

The first-line supervisor is critical to the management of the employer's safety and health effort. The reasoning for such a statement is vested in the function of the first-line supervisor, who is on the front line.

Safety and Health (Managing)

Safety and health starts in the design of the building. If the applicable safety and health codes are not an integral part of the design, then you are stymied from the start. Almost every design component has a safety factor attached to it. It might be what force the building would withstand, the fire protection within the building, the adequacies of egress, the provision of proper ventilation, or the overall habitability of the structure. These are some of the large items. Many small subtle features go into the design of office buildings (e.g., exit signage) that provide for safety and health. Other items that are not part of the building's structural design but are an integral part of the feeling of being safe would be the design and wiring of the security system for the facility.

Once a facility has been designed with safety in mind, then planning for safe and healthy occupancy must be considered. This may include everything from hygiene facilities to the amount of space per worker. Owners and employers must work together to assure that the actual workplaces are as free from hazards as possible. We are looking not only at those factors that could cause physical injury but also at those that could produce acute or chronic health effects from short- or long-term exposure to such items as mold or wastewater.

This facility or worksite is the employer's and workers' home away from home. Thus, it may need to be more livable than the actual workers' homes since they spend most of their waking hours in office environments. Study after study indicates that in most cases, a good (safe and healthy) work environment goes a long way toward fostering job satisfaction, high morale, and better productivity. At least, this is the foundation for a productive workplace. On Dr. Abraham Maslow's hierarchy of needs, once the basic biological needs of food and shelter are met, security is the second most important need. Thus, providing a safe and healthy work environment is a giant step toward meeting this need of workers.

Once the physical environment is safe and healthy, does anyone really care about the workers' safety and health? This is where managing safety and health comes into play. It is not enough to espouse worker safety and health. It is imperative that the employer is committed to providing the leadership regarding OSH.

Employers need to do more than talk a good talk, since their actions will speak louder than their words. Safety and health needs to be part of the culture of the workplace. The employer and the workers should not tolerate an unsafe or unhealthy workplace.

Management of safety and health not only requires commitment and leadership support as a critical element. Management of OSH demonstrates commitment to the effort, as does having designated resources and a separate budget for safety and health management. Employee involvement in fostering safety and health has been shown to be another key factor. As part of the management process, hazard identification, hazard intervention and prevention, and training must exist. Training is vital even in the work environment since workers need to know the company's policies and rules, the hazards they are going to be exposed to, how to use equipment safely, what to do in an emergency, and their responsibilities related to safety and health at the company.

There is a real cost when owners and employers do not take safety and health seriously. Workers' compensation is integrally tied to workplace injuries and illnesses. It is a direct cost to a business when workers become ill or suffer an injury in their workplace. The more injuries, the more the cost the employer will experience. The workers compensation system demonstrates the importance of prevention as a mechanism to hold down the cost of workers' compensation. The containment of workers compensation premiums is critical to a profitable business operation, whether it be heavy industry or office-related work.

Why Is Managing Safety and Health a Needed Entity?

The following principles illustrate why safety and health needs to be managed:

1. An unsafe act, an unsafe condition, and an accident or incident are all indicators or symptoms of something wrong in the management system. Many factors contribute to an accident or incident, so it is imperative that all contributing factors are considered in order to determine their underlying causes.

2. We can predict that certain sets of circumstances have the potential to produces severe injuries. These circumstances can be identified and controlled. Thus, it is possible to predict severity when certain conditions exist. This will allow for resources to be directed toward the more frequently occurring accidents/incidents that are the riskiest or most severe versus the less severe or less frequently occurring types of accidents.

   History has shown that situations that involve more severe or catastrophic injuries/incidents are reasonably predictable, for example,

   a. Unusual, nonroutine work

   b. Nonproduction activities such as maintenance

   c. Sources of high energy

   d. Certain construction situations such as steel erection, tunneling, working over water, or handling explosives

   e. Lifting or material-handling tasks

   f. Repetitive motion situations

   g. Psychological stress situations

   h. Occupations with dangerous exposures to hazards

   i. Exposure to toxic materials

3. Safety should be managed like any other company function. Management should direct the safety and health effort by setting achievable goals using management tools such as planning, organizing, and controlling techniques to facilitate their accomplishment. Safety and health must be a functional element of management.

4. The key to effective line safety and health performance is the use of management procedures that fix accountability and hold accountable those responsible for safety and health as a key element of their job performance.

   Any line manager will usually achieve results in those areas where they are being measured by management. In most all cases, someone who is held accountable will take responsibility. Accountability fosters responsibility while the lack of accountability results in failure to be responsible.

5. The function of safety and health is to locate and define the operational errors that allow accidents/incidents to occur. The detection of such errors helps by asking why certain accidents/incidents happen. This is often defined as searching for the root causes. This entails that the safety and health professional must look beyond direct causes (release of energy) and indirect causes such as acts or conditions to basic causes, which entails management policies and procedures, personal practices or factors, and environmental factors.

By ascertaining whether certain known effective controls are being utilized, the symptom (act, condition, accident, or incident) can be traced back to see why it was allowed to occur. Looking at the company's system (procedures) and asking whether certain things are done in a predetermined manner that is known to be successful.

Some questions that can be asked as to why (or why not) certain known effective controls are being utilized are as follows:

a.   What is management safety and health policy?

b.   How is the company organized?

c.   What is the function of the safety and health department?

d.   What is safety and health's niche in the organization?

e.   What are the relationships between the production and line staff?

6. The cause of unsafe or risky behavior can be identified and classified, which provides logical reasons as to why individuals react the way that they do. At times, employees are placed in overload situations that require them to experience heavier workload beyond not only their physical abilities but also their mental capacities. These are bona fide stressors.

   Other times, workers feel trapped and make mistakes because the assigned task or situation is beyond their skill level, again both physically and mentally; also, the lack of training and experience with the task come into play. The pressure to accomplish the assigned task to save face and their job results in the potential for error.

   A less understood reason why workers become victims of an accident/incident is their own decision to commit an error because work must get done, which involves a degree of logic in choosing the unsafe act. After all, the belief is that accidents/incidents happen to other people. This type of motivation makes a lot more sense to the workers' current mental thought process. It causes them to make the decision to operate unsafely. It is a mental decision based on their perception of reality and their mental condition at that moment.

   Management's task with respect to safety and health is to identify and deal with the causes of unsafe behaviors, their cause, effect, and prevention, and not the behavior itself.

7. In most cases, unsafe behavior is normal behavior; it is the result of normal people reacting to the environment of the workplace. Management's responsibility is to change the environment that has led to the unsafe behavior. In most cases, unsafe behavior is the result of an environment that has been constructed or designed by management.

8. There are three major subsystems that must be dealt with in building an effective safety and health system. They are as follows:

a.   The physical

b.   The managerial

c.   The behavioral

   This entails the analysis of issues, the development of systems of control, and communicating those systems to the line organization and monitoring the results achieved.

9. The safety and health system should fit the culture of the organization. The way to manage safety and health must be consistent with existing culture and must be flexible enough to change to be consistent with other functions within the workplace.

10. There is no one right or correct way to achieve safety and health in a company or business; however, for a safety and health system to be effective, it must meet certain criteria. These criteria are as follows:

    a. Force supervisory performance

    b. Involve middle management

    c. Have top management visibly showing their commitment

    d. Have employee participation

    e. Be flexible

    f. Be perceived as positive

Summary: Why Management?

A short summary as to why management is needed to accomplish the desired outcome for the OSH initiative is as follows:

1. Management is ultimately responsible for OSH.

   Thus, the need for commitment, budgeting, and planning for safety falls upon management's shoulders.

2. Poor safety conditions and safety performance by the workforce result from management's failure to effectively manage workplace safety and health.

   Accidents and incidents result from management's inability to manage safety and health as they would any other company function.

3. Worker and supervisor involvement are critical to good workplace safety and health.

   A workforce that is not involved in safety and health has no ownership and thus feels no investment, responsibility, or accountability for it.

4. Workplace safety and health is not a dynamic, fast-evolving component of the workplace, since it should go hand in hand with your normally evolving business.

   In safety and health, there is little that is new, since we know the causes of occupational injuries and illnesses as well as how to intervene to mitigate and prevent their occurrence. There should be no excuses for accidents and incidents, since the philosophy should be that all are preventable.

5. You cannot have an effective safety and health program without specifically holding the first-line supervisors accountable for their own and their employees' safety and health performance (as well as other management personnel).

   First-line supervisors are the key to the success or failure of a safety and health program. All your planning, budgeting, and goal setting are for naught if they are not accountable and committed to safety and health.

6. Hazard identification and analysis are critical functions in assuring a safe and healthy work environment.

   If management and the workforce do not tolerate the existence of hazards and constantly ask the question, "How could this have happened?", they are better able to get to the basic causes of adverse workplace events.

7. Management's philosophy, actions, policies, and procedures regarding safety and health in the workplace put workers into situations where they must disregard good safety and health practices in order to perform their assigned task or work.

   Workers in most cases perform work in an unsafe or unhealthy manner when they have no choice or are forced to do so by existing conditions and expectations.

8. It is critical to obtain safe and healthy performance or behavior by effective communications and motivational procedures that are compatible with the culture of your workforce.

   If you do not understand what it is that fulfills the needs of your workforce, you will not be able to communicate or motivate them regarding safety and health outcomes no matter how good your management approach.

Further Readings

Reese, C.D. *Accident/Incident Prevention Techniques (Second Edition)*. Boca Raton, FL: CRC Press, 2012.

Reese, C.D. *Occupational Health and Safety Management (Third Edition)*. Boca Raton, FL: CRC Press, 2016.

Reese, C.D. and J.V. Eidson. *Handbook of OSHA Construction Safety & Health (Second Edition)*. Boca Raton, FL: CRC/Lewis Publishers, 2006.

7

Safety and Health Programs

The *why* of having a safety and health program takes some degree of explaining since many individuals and employers view such an endeavor as a paper game or just a waste of paper. It also takes time for the employer to put his/her safety and health program in writing. The following two statements indicate why a written safety and health program is a desirable undertaking by an employer.

- A statement of policy, purpose, and expectations that the company has for safety and health helps to set the foundation for the effort.
- It is recognized the companies with a written safety and health program have fewer accidents or incidents that result in job-related injuries and illnesses.

The need for safety and health programs in the workplace has been an area of controversy for some time. Many companies feel that written safety and health programs are just more paperwork, a deterrent to productivity, and nothing more than another bureaucratic way of mandating safety and health on the job. But over a period of years, data and information have been mounting in support of the need to develop and implement written safety and health programs.

In order to effectively manage safety and health, a company must pay attention to some critical factors. These factors are of the essence in managing safety and health on work sites. The questions that need to be answered regarding managing safety and health are as follows:

1. What is the policy of management regarding safety and health at the company's workplace?
2. What are the safety and health goals for the company?
3. Who is responsible for OSH?
4. How are supervisors and employees held accountable for job safety and health?
5. What are the safety and health rules for this type of industry?
6. What are the consequences of not following the safety rules?
7. Are there set procedures for addressing safety and health at the work site?

Studies conducted indicate that companies with written safety and health programs are shown to have more accidents if they

- Do not have a separate budget for safety
- Do not have training for new hires
- Have no outside source for safety training
- Have no specific training for supervisors
- Do not conduct inspections

- Are using canned programs
- Have no written program
- Have no employee safety committee
- Have no membership in professional safety organization
- Have no mechanism to recognize safety accomplishment
- Do not document or review accidents as part of safety
- Do not hold supervisors accountable for safety as part of their job
- Do not have top management actively promoting safety awareness

It seems apparent from the previous results of research that in order to have an effective safety program, at a minimum, an employer must

- Have a demonstrated commitment to job safety and health
- Commit budgetary resources
- Train new personnel
- Insure that supervisors are trained
- Have a written safety and health program
- Hold supervisors accountable for safety and health
- Respond to safety complaints and investigate accidents
- Conduct safety audits

Other refinements can always be part of the safety and health program, which will help in reducing workplace injuries and illnesses. They are as follows: more worker involvement (for example, joint labor/management committees); incentive or recognition programs; getting outside help from a consultant or safety association; and setting safety and health goals. A decrease in occupational incidents that result in injury, illness, or damage to property is enough reason to develop and implement a written safety and health program.

Why Have a Comprehensive Safety and Health Program?

The three major considerations involved in the development of an occupational safety and health program are as follows:

1. **Humanitarian.**

    Safe operation of workplaces is a moral obligation imposed by modern society. This obligation includes consideration for loss of life, human pain and suffering, family suffering, hardships, etc.

2. **Legal obligation.**

    Federal and state governments have laws charging the employer with the responsibility for safe working conditions and adequate supervision of work practices. Employers are also responsible for paying the costs incurred for injuries suffered by their employees during their work activities.

3. **Economic.**

Prevention costs less than accidents. This fact is proven consistently by the experience of thousands of industrial operations. The direct cost is represented by medical care, compensation, etc. The indirect cost of 4 to 10 times the direct cost must be calculated, as well as the loss of wages to employees and the reflection of these losses on the entire community.

All three of these are good reasons to have a safety and health program. It is also important that these programs be formalized in writing, since a written program sets the foundation and provides a consistent approach to occupational safety and health for the company. There are other logical reasons for a written safety and health program. Some of them are as follows:

- It provides standard directions, policies, and procedures for all company personnel.
- It states specifics regarding safety and health and clarifies misconceptions.
- It delineates the goals and objectives regarding workplace safety and health.
- It forces the company to actually define its view of safety and health.
- It sets out in black and white the rules and procedures for safety and health that everyone in the company must follow.
- It is a plan that shows how all aspects of the company's safety and health initiative work together.
- It is a primary tool of communication of the standards set by the company regarding safety and health.

Why Build an Occupational Safety and Health Program?

The length of such a written plan is not as important as the content. It should be tailored to the company's needs and the safety and health of its workforce. It could be just a couple of pages or a many paged document. In order to insure a successful safety program, three conditions must exist. These are as follows: management leadership, safe working conditions, and safe work habits by all employees. The employer must

- Let the employees know that he/she is interested in safety on the job by consistently enforcing and reinforcing safety regulations.
- Provide a safe working place for all employees; it pays dividends.
- Be familiar with federal and state laws applying to his/her operation.
- Investigate and report all Occupational Safety and Health Administration (OSHA) recordable accidents and injuries. This information may be useful in determining areas where more work is needed to prevent such accidents in the future.
- Make training and information available to employees, especially in such areas as first aid, equipment operation, and common safety policies.
- Develop a prescribed set of safety rules to follow, and see that all employees are aware of the rules.

The basic premise is all employers should establish a workplace safety and health program to assist them in compliance with OSHA standards and the General Duty Clause of the Occupational Safety and Health Act of 1970 (OSHAct) [Section 5(a)(1)]. Each employer should set up a safety and health program to manage workplace safety and health to reduce injuries, illnesses, and fatalities by a systematic approach to safety and health. The program should be appropriate to conditions in the workplace, such as the hazards to which employees are exposed and the number of employees there. The primary guidelines for employers to develop an organized safety and health program are as follows:

1. Employers are advised and encouraged to institute and maintain in their establishments a program that provides systematic policies, procedures, and practices that are adequate to recognize and protect their employees from OSH hazards.

2. An effective program includes provisions for the systematic identification, evaluation, and prevention or control of general workplace hazards, specific job hazards, and potential hazards that may arise from foreseeable conditions.

3. Although compliance with the law, including specific OSHA standards, is an important objective, an effective program looks beyond specific requirements of law to address all hazards. This effectively will seek to prevent injuries and illnesses, whether or not compliance is an issue.

4. The extent to which the program is described in writing is less important than how effective it is in practice. As the size of a work site or the complexity of a hazardous operation increases, however, the need for written guidance increases to ensure clear communication of policies and priorities and consistent and fair application of rules.

The primary elements that should be addressed within this program are management leadership and employee participation; hazard identification and assessment; hazard prevention and control; information and training; and evaluation of program effectiveness according to OSHA guidelines.

A review of research on successful safety and health programs reveals a number of factors that these programs comprise. Strong management commitment to safety and health and frequent, close contacts between workers, supervisors, and management on safety and health are the two most dominant factors in good safety and health programs. Other relevant factors include workforce stability, stringent housekeeping, training emphasizing early indoctrination and follow-up instruction, and special adaptation of conventional safety and health practices to enhance their suitability to the workplace.

Components of a Safety and Health Program

A listing of the potential components that a successful safety and health program comprises is as follows:

1. Safety and health program management
2. Inspections and job observations

3. Illness and injury investigations

4. Task analysis

5. Training

6. Personal protection

7. Communication/promotion of safety and health

8. Personal perception

9. Off-the-job safety and health

This is only a representative list that could be either expanded or consolidated depending upon the unique needs of the company/contractor. Safety and health programs should be tailored to meet individual requirements.

Evaluative Questions Regarding Safety and Health Programs

This section will break down these components into subparts or subelements by using a questioning approach as the mechanism to draw attention to the intricacies of a safety and health program.

These breakdown and questions, which should guide your evaluation and/or development of a successful safety and health program, are as follows:

1. Safety and Health Program Management
   a. Is there a safety policy signed and dated by top management?
   b. Is there someone responsible for safety and health?
   c. Does a safety and health manual or handbook exist?
   d. Is there a set time devoted to safety and health during management meetings?
   e. Are employees encouraged to participate in the safety and health program?
   f. Are there safety and health rules and regulations for employees and specific jobs?
   g. Is there a discipline policy for disobeying safety and health rules?
   h. Are safety and health rules enforced and violators disciplined?
   i. Who is held accountable for safety and health?
   j. Are there special goals for safety and health?
2. Inspections/Job Observations
   a. Inspections
      i. Are safety and health inspections conducted?
      ii. How often are inspections conducted?
      iii. Are unsafe conditions or hazards found and corrected immediately?
      iv. Is equipment inspected?
      v. Is equipment maintained?
      vi. Is there a preventive maintenance program?

      vii. Are housekeeping inspections conducted?

     viii. Is housekeeping maintained?

      ix. Does monitoring occur for health hazards?

       x. Are written inspection reports completed?

      xi. Are inspection reports disseminated?

  b. Job observations

       i. Are job observations done?

      ii. Who does job observations?

     iii. Do job observations result in new work practices, workplace design, training, retraining, tasks analysis, Job Safety Analysis (JSA), or Safe Operating Procedures (SOPs)?

     iv. Are job observations done for punitive reasons?

3. Illness and Injury Investigations

  a. Are employees encouraged to report hazards and accidents?

  b. Are all incidents involving illness or injury investigated?

  c. Have written reports been generated for all incidents?

  d. Are preventive recommendations being implemented?

  e. Do employees review incident reports?

  f. Are incident data analyzed to determine illness or injury trends and damage experience?

4. Task Analysis

  a. Do inspections, job observations, and incident investigations result in a task analysis?

  b. Do task analyses result in changes in work practices or workplace design?

  c. Does task analysis facilitate the development of JSAs or SOPs?

  d. Does task analysis result in new training or retraining?

5. Training

  a. Do all employees receive safety and health training?

  b. Do employees receive site-specific training?

  c. Are employees given job-specific or task-specific training?

  d. Is training well planned and organized?

  e. Do both classroom and on-the-job training (OJT) or job instruction training (JIT) training occur?

  f. Is refresher training conducted?

  g. Do management and supervisors receive safety and health training?

  h. Are training records maintained?

6. Personal Protection

  a. Does the work require personal protective equipment (PPE)?

  b. Is the proper PPE available?

    c. Have employees been trained in the use of PPE?

    d. Do you need a respirator program (29 CFR 1910.134)?

    e. Should you follow the requirements for handling hazardous waste (29 CFR 1910.120)?

    f. Are the rules and use of PPE enforced?

7. Communication/Promotion of Safety and Health

    a. Communication

        i. Is safety and health visible?

        ii. Are company/contractor safety and health goals communicated?

        iii. Are safety and health meetings held (i.e., toolbox)?

        iv. Do safety and health talks convey relevant information?

        v. Are personal safety and health contacts made?

        vi. Are bulletin boards used to communicate safety and health issues?

        vii. Do those responsible for safety and health request feedback?

        viii. Are safety and health suggestions given consideration and/or used?

        ix. Are supervisors interested in safety and health?

    b. Promotion

        i. Is there an award/incentive program tied to safety and health?

        ii. Are safety and health exhibits or posters used?

        iii. Are paycheck stuffers used?

8. Personal Perception

    a. Does the company/contractor extend considerable effort to assure an effective safety and health program?

    b. Do supervisors support and enforce all aspects of the safety and health program?

    c. Do most employees insist on doing all tasks in a safe and healthy manner?

9. Off-the-Job Safety and Health

    a. Is off-the-job safety and health promoted as part of the total safety and health program?

    b. Does the company/contractor provide a wellness/fitness program?

    c. Does the company/contractor foster and encourage healthier lifestyles?

    d. Does the company/contractor support an employee assistance program?

Tools for a Safety and Health Program Assessment

There are three basic methods for assessing safety and health program effectiveness. This discussion will explain each of them. It also will provide more detailed information on how to use these tools to evaluate each element and subsidiary component of a safety and health program. The three basic methods for assessing safety and health program effectiveness are as follows:

1. Checking documentation of activity
2. Interviewing employees at all levels for knowledge, awareness, and perceptions
3. Reviewing site conditions and, where hazards are found, finding the weaknesses in management systems that allowed the hazards to occur or to be "uncontrolled."

Some elements of the safety and health program are best assessed using one of these methods. Others lend themselves to assessment by two or all three methods.

A key to formalizing the management of safety and health in a workplace is the safety and health program. Many feel that written safety and health programs are just more paperwork, a deterrent to productivity, and nothing more than another bureaucratic way of mandating safety and health on the job. But over a period of years, data and information have been mounting in support of the need to develop and implement written safety and health programs for all workplaces.

This perceived need for written programs must be tempered with a view to their practical development and implementation. A very small employer who employs one to four employees and no supervisors in all likelihood needs only a very basic written plan, along with any other written programs that are required as part of an OSHA regulation. But, in large environments where the number of employees increases, the owner/employer becomes more removed from the hands-on aspects of what now may be multiple floor complexes or different types of work sites in the workplace.

Now you must find a way to convey support for safety to all those who work in the same facility. As with all other aspects of business, the employer must plan, set the policies, apply management principles, and assure adherence to the goals in order to facilitate the efficient and effective completion of projects or work. Again, job safety and health should be managed the same as any other part of the office building's business.

The previous paragraph simply states that in order to effectively manage safety and health, an owner/employer must pay attention to some critical factors. These factors are of the essence in managing safety and health on work sites. The questions that need to be answered regarding managing safety and health are as follows:

1. What is the policy of the owner/employer regarding safety and health in the workplace?
2. What are the safety and health goals for the owner/employer?
3. Who is responsible for OSH?
4. How are supervisors and employees held accountable for job safety and health?
5. What are the safety and health rules for the office building environment?
6. What are the consequences of not following the safety rules?
7. Are there set procedures for addressing safety and health issues that arise in the workplace?
8. How are hazards identified?
9. How are hazards controlled or prevented?
10. What type of safety and health training occurs? And, who is trained?

Specific actions can be taken to address each of the previous questions. The written safety and health program is of primary importance in addressing these items.

Written safety and health programs have a real place in modern safety and health practices, not to mention the potential benefits. If a decrease in occupational incidents that result in injury, illness, or damage to property is not reason enough to develop and implement a written safety and health program, the other benefits from having a formal safety and health program seem well worth the investment of time and resources. Some of these are as follows:

- Reduction of industrial insurance premiums/costs
- Reduction of indirect costs of accidents
- Fewer compliance inspections and penalties
- Avoidance of adverse publicity from deaths or major accidents
- Less litigation and fewer legal settlements
- Lower employee payroll deductions for industrial insurance
- Less pain and suffering for injured workers
- Fewer long-term or permanent disability cases
- Increased potential for retrospective rating refunds
- Increased acceptance of bids on more jobs
- Improved morale and loyalty from individual workers
- Increased productivity from work crews
- Increased pride in company personnel
- Greater potential of success for incentive programs

Why Other Required Written Programs?

Although there is not a regulatory requirement for businesses, corporations, or companies to have written OSH programs, many of the OSHA regulations have requirements for written programs that coincide with the regulations. This may become a bothersome requirement to many within the workplace, but the failure to have these programs in place and in writing is a violation of the regulations and will result in a citation for the company. At times, it is difficult to determine which regulations require a written program, but in most cases, the requirements are well known. Some of the other OSHA regulations that require written programs are as follows:

- Blood-borne pathogens/exposure control plan
- Confined space entry program
- Crane and derrick safety compliance program
- Drug-free workplace compliance program
- Emergency action compliance program
- Fall protection compliance program
- Electrical safety compliance program
- Fire prevention compliance program
- Hazard communications program

- Hazardous waste compliance program
- Hearing conservation compliance program
- Injury and illness prevention compliance program
- Laboratory safety compliance program
- Lead safety compliance program
- Lockout/tagout/energy control program
- Machine safeguarding compliance program
- Medical surveillance compliance program
- Motor vehicle occupant compliance program
- Personal protective compliance program
- Process safety compliance program
- Respiratory protection compliance program
- Safety and health compliance program

The specific requirements for the content of written programs vary with the regulation.

Summary

The management of safety and health is well recognized as a vital component by those who have responsibility for workplace safety and health. It is not a just a written proclamation or program but a true and supported endeavor to provide a safe and healthy workplace for workers. It must be as well planned and organized as any other facet of the company's business. Managing safety and health is critical to be able to provide the protections that a workforce within a workplace is entitled to and deserves. This management process will probably go much better if a person, who by training and/or experience, can take the responsibility for the development of the safety and health effort in the work environment; someone must do it if there is to be a safe and healthy environment. Thus, workplace safety must be planned, organized, and implemented in a businesslike manner in order for it to be successful in assuring that an office building is safe, healthy, and secure.

Further Readings

Reese, C.D. *Mine Safety and Health for Small Surface Sand/Gravel/Stone Operations: A Guide for Operators and Miners.* Storrs, CT: University of Connecticut Press, 1997.

Reese, C.D. *Accident/Incident Prevention Techniques (Second Edition).* Boca Raton, FL: CRC Press, 2012.

Reese, C.D. *Occupational Health and Safety Management (Third Edition).* Boca Raton, FL: CRC Press, 2016.

Reese, C.D. and J.V. Eidson. *Handbook of OSHA Construction Safety & Health (Second Edition).* Boca Raton, FL: CRC/Lewis Publishers, 2006.

United States Department of Labor. Occupational Safety and Health Administration. *Office of Construction and Maritime Compliance Assistance: OSHA Instruction STD 3-1.1*. June 22, 1987.

United States Department of Labor. Occupational Safety and Health Administration. *Federal Register: Safety and Health Program Management Guidelines*. Vol. 54, No. 16, pp. 3904–3916. January 26, 1989.

United States Department of Labor. Occupational Safety and Health Administration. *Field Inspection Reference Manual (FIRM): OSHA Instruction CPL 2.103*. Washington, DC: US Department of Labor, September 26, 1994.

United States Department of Labor. Occupational Safety and Health Administration. *Citation Policy for Paperwork and Written Program Requirement Violations: OSHA Instruction CPL 2.111*. Washington, DC: US Department of Labor, November 27, 1995.

8

Special Emphasis Programs

Special emphasis programs are another of the tools used in safety and health as an effective accident prevention technique. Any time a special program is instituted that targets a unique safety and health issue, an organized approach in prevention needs to be developed. The benefits of instituting a special program include the fact that the potential hazard is kept on everybody's mind, management receives feedback, and workers receive reinforcement for the desired performance. A program can be developed in any area where the company feels there is a need. Some areas of focus could be as follows: ladder safety, back injuries, health-related issues, vehicle or equipment safety, power tool incidents, etc. For success, the program may contain goals to attain, rewards to receive, or even consequences for enforcement, if the rules of the program are not followed. By setting up a program, the company is at least taking action to target accidents and prevent their occurrence.

Why are special emphasis programs necessary?

- Special emphasis is needed for unique problems.
- They can place their emphasis on occupational safety and health (OSH).
- They can target any area in the workplace needing special attention.
- They develop processes and procedures to address pressing safety and health issues.
- Scarce resources are targeted to real problems, not a scattered approach.
- They allow the development of targeted/specific materials or techniques.
- They are used for comprehensive approaches to solving safety and health problems.

Summary

Special emphasis programs are designed to meet the unique OSH needs of the company. An example of a special emphasis ladder safety program could just as easily be an accident repeat program for those individuals who have a propensity to suffer from recurring occupational accidents. These types of programs may be very simple or very complex depending upon the seriousness, number of occurrences, or amount of resources that one desires to invest. The special emphasis program definitely has its place in accident prevention initiatives.

Further Readings

Reese, C.D. *Accident/Incident Prevention Techniques (Second Edition)*. Boca Raton, FL: CRC Press, 2012.

Reese, C.D. *Occupational Health and Safety Management (Third Edition)*. Boca Raton, FL: CRC Press, 2016.

Reese, C.D. and J.V. Eidson. *Handbook of OSHA Construction Safety & Health (Second Edition)*. Boca Raton, FL: CRC/Lewis Publishers, 2006.

9

Accident Investigations

Millions of accidents and incidents occur throughout the United States every year. The inability of people, equipment, supplies, or surroundings to behave or react as expected causes most of the accidents and incidents. Accident and incident investigations determine how and why each incident occurs. By using the information gained through an accident and incident investigation, a similar or perhaps more disastrous accident may be prevented. Accident and incident investigations should be conducted with prevention in mind. The mission is one of fact finding. Investigations are not to find fault.

An accident, by definition, is any unplanned event that results in personal injury or in property damage. When the personal injury requires little or no treatment, or is minor, it is often called a first-aid case. If it results in a fatality or in a permanent total, permanent partial, or temporary total (lost-time) disability, it is serious. Likewise, if property damage results, the event may be minor or serious. All accidents/incidents should be investigated regardless of the extent of injury or damage.

Accidents are part of a broad group of events that adversely affect the completion of a task. Accidents fall under the category of an incident. With this said, the most commonly used term for accidents and incidents is *accident*, which is used to refer to both accidents and incidents, since the basic precepts are applicable to both.

An important element of a safety and health program is accident investigation. Although it may seem to be too little too late, accident investigations serve to correct the problems that contribute to an accident and will reveal accident causes that might otherwise remain uncorrected.

The reasons why accident investigations are an important element within a company's safety and health initiative are as follows:

- To determine cause and effect
- To systematically collect and gather information on the incident
- To determine the facts, not the faults
- To provide data for comparing and contrasting
- To take short- and long-term prevention actions
- To provide an overview of success and failure of safety and health
- To help others in solving and preventing accidents/incidents in their operations
- To standardize the investigational process
- To be an integral part of the occupational safety and health (OSH) initiative

The main purpose of conducting an accident investigation is to prevent a recurrence of the same or a similar event. It is important to investigate all accidents regardless of the extent of injury or damage. The kinds of accidents that should be investigated and reported are as follows:

- Disabling injury accidents
- Nondisabling injury accidents that require medical treatment
- Circumstances that have contributed to acute or chronic occupational illness
- Noninjury, property damage accidents that exceed a normally expected operating cost
- Near accidents (sometimes called near miss) with a potential for serious injury or property damage

In spite of their complexity, most accidents are preventable by eliminating one or more causes. Accident investigations determine not only what happened but also how and why. The information gained from these investigations can prevent recurrence of similar or perhaps more disastrous accidents. Accident investigators are interested in each event as well as in the sequence of events that led to an accident. The accident type is also important to the investigator. The recurrence of accidents of a particular type or those with common causes shows areas needing special accident prevention emphasis.

It is important to have some mechanism in place to investigate accidents and incidents in order to determine the basis of cause-and-effect relationships. You may determine these types of relationships only when you actively investigate all accidents and incidents that result in injuries, illnesses, or damage to property, equipment, and machinery.

Accident investigation becomes more effective when all levels of management, particularly top management, take a personal interest in controlling accidents. Management adds a contribution when it actively supports accident investigations. It is normally the responsibility of line supervisors to investigate all accidents; in cases where there is serious injury or equipment damage, other personnel such as department managers and an investigation team might become involved as well.

Once types of accidents or incidents have been determined, then prevention and intervention activities can be undertaken to assure that those types do not reoccur. Even if the company is not experiencing large numbers of accidents and incidents, it still needs to implement activities that actively search for, identify, and correct the risk from hazards on jobsites. Reasons to investigate accidents and incidents include the following:

- To know and understand what happened
- To gather information and data for present and future use
- To determine cause and effect
- To provide answers for the effectiveness of intervention and prevention approaches
- To document the circumstances for legal and workers' compensation issues
- To become a vital component of the safety and health program.

If the company has only a few accidents and incidents, it might want to move down one step to examine near misses and first aid-related cases. It is only a matter of luck or timing that separates the near miss or first-aid event from being a serious, recordable, or reportable event. The truth is that the company probably has been lucky by seconds or inches. (A second later and a tool would have hit someone, or an inch more and it would have cut off a finger.) Truly, it pays dividends to take time to investigate accidents and incidents occurring in the workplace.

Reporting Accidents

When accidents are not reported, their causes usually go uncorrected, allowing the chance for the same accident to result again. Every accident, if properly investigated, serves as a learning experience to the people involved. The investigation should avoid becoming a mechanical routine. It should strive to establish what happened, why it happened, and what must be done to prevent a recurrence. An accident investigation must be conducted to find out the facts and not to place blame.

There are sound reasons for reporting accidents, such as the following:

- You learn nothing from unreported accidents.
- Accident causes go uncorrected.
- Infection and injury aggravations can result.
- Failure to report injuries tends to spread and become an accepted practice.

The results-oriented supervisor recognizes that the real value of investigation can only be achieved when his/her workers report every problem, incident, or accident that they know of. In order to promote conscientious reporting, it may be helpful to know some of the reasons why workers fail to, or avoid, reporting accidents. There are usually reasons that workers espouse for not reporting accidents:

1. Fear of discipline
2. Concern about the company's record
3. Concern for their reputation
4. Fear of medical treatment
5. Dislike of medical personnel
6. Desire to keep personal record clear
7. Avoidance of red tape
8. Desire to prevent work interruptions
9. Concern about attitudes of others
10. Poor understanding of importance

How can a company combat these reporting problems?

1. React in a more positive way.
2. Indoctrinate workers on the importance of reporting all accidents.
3. Make sure everyone knows what kinds of accidents should be reported.
4. Give more attention to prevention and control.
5. Recognize individual performance.
6. Develop the value of reporting.
7. Show disapproval of injuries neglected and not reported.
8. Demonstrate belief by action.
9. Do not make mountains out of molehills.

Let the worker know that the company appreciates his/her reporting promptly. Inquire about his/her knowledge of the accident. Don't interrogate or grill the person. Stress the value of knowing about problems while they are still small. Focus on accident prevention and loss control. Emphasize compliance with practices, rules, and protective equipment, and promptly commend good performance. Pay attention to the positive things workers do, and give sincere, meaningful recognition where and when it is deserved. Use compliments as often as warnings are used. Use group and personal meetings to point out and pass on knowledge gained from past accidents. Give an accident example as an important part of every job instruction. Show belief in what the company officials say by taking corrective action promptly. Something can always be done at that moment, even if permanent correction requires time to develop new methods, buy new equipment, or modify the building.

The first step in an effective investigation is the prompt reporting of accidents. A company can't respond to accidents, evaluate their potential, and investigate them if they are not reported when they happen. Prompt reporting is the key to effective accident investigations. Hiding small accidents doesn't help prevent the serious accidents that kill people, put the company out of business, and take away jobs. If workers don't report accidents to the supervisor, they are stealing part of the supervisor's authority to manage his/her job.

Line supervisors should be involved in accident investigations, basically, because they are normally the people in direct contact with the worker, and understand their problems, personalities, and capabilities. Involving supervisors increases their responsibilities toward the accident prevention effort. Supervisors are normally responsible for training, so becoming involved in investigations will make them more aware of what causes accidents as well as ways to prevent a recurrence. There are benefits from having a line supervisor involved in accident investigations.

Accident investigation should be preplanned by developing a formalized investigative process. It should encompass the tools needed to conduct the investigation, the procedural steps should be planned out, and a written record should be developed and maintained that answers the questions *who, what, when, where, how, with what, why, with whom,* and *how much.* The process should be documented using photographs and visuals. All these items are best accomplished using an investigative team approach. This team will also conduct and document all interviews. The team is to be charged with collecting and securing all evidence that will be used to develop the final report of the team.

Summary

Once the accident investigation process is complete and the final report done, a review process should be in place, with management involvement, to determine the actions to be taken regarding the findings and recommendations from the final report. A process should be installed or formulated to ensure the implementation of interventions and preventive activities. Once the implementation process is in place and active, its overall effectiveness needs to be determined. Companies will want to know how the accident investigation process and outcome impacts their safety and health program and allows them to assess the need for change in their safety and health effort. Accident investigations are an integral part of any good safety and health initiative or program. Investigations should be performed for all accidents, injuries, illnesses, property damage, and near-miss incidents. It should be a formalized part of a company's safety and health commitment.

Further Readings

Reese, C.D. *Accident/Incident Prevention Techniques (Second Edition).* Boca Raton, FL: CRC Press, 2012.

Reese, C.D. *Occupational Health and Safety Management (Third Edition).* Boca Raton, FL: CRC Press, 2016.

Reese, C.D. and J.V. Eidson. *Handbook of OSHA Construction Safety & Health (Second Edition).* Boca Raton, FL: CRC/Lewis Publishers, 2006.

United State Department of Labor, Mines Safety and Health Administration. *Accident Investigation (Safety Manual No. 10).* Beckley: Revised 1990.

United States Department of Labor, National Mine Health and Safety Academy. *Accident Prevention Techniques: Accident Investigation.* Beckley: 1984.

United States Department of Labor, Occupational Safety and Health Administration, Office of Training and Education. *OSHA Voluntary Compliance Outreach Program: Instructors Reference Manual.* Des Plaines IL: US Department of Labor, 1993.

10

Training

The reason for addressing education and training as an integral part of the occupational safety and health (OSH) approach by a company is to be assured of the qualifications of managers, supervisors, and employees by training them in the expectations, behaviors, and commitment to safety and health by the employer. This is why support of training and education for the workforce has its importance to OSH. A well-educated and well-trained workforce results in better control of safety and health outcomes.

Why train employees regarding OSH? Any time an employee does not have the skill or skills needed to perform a job in a safe manner, he/she needs to be trained. The failure to have a trained workforce will have an impact upon the following:

- Productivity
- Motivation
- Cost
- Numbers of accidents/incidents

Other factors as to why safety and health education and training are important to safety and health endeavors are as follows:

- To be able to perform all work-related tasks in safe and healthy manner
- To set forth the employer's OSH expectations
- To allow the workers to protect themselves and understand how to avoid the hazards of their jobs

When to Train

There are appropriate times when safety and health training should be provided. They are when

- A worker lacks the safety skills
- Occupational Safety and Health Administration (OSHA) regulation requires or mandates training
- An employee is promoted to supervisor
- A new employee is hired
- An employee is transferred to another job or task
- New equipment, machinery, or vehicles are brought into the workplace

- Changes have been made in the normal operating procedures
- A worker has not performed a task for some period of time
- It is required by OSH regulations

If a worker could not perform the task of his/her job safely if his/her life depended upon it, then training would provide the skills for him/her to do so. Thus, if a worker could safely perform his/her job if it were a matter of life or death, then the worker has a behavioral problem and not a skill issue.

Safety and health training is critical to achieving accident prevention. Companies that do not provide new-hire training, supervisory training, and worker safety and health training have an appreciably greater number of injuries and illnesses than companies who carry out safety and health training

It has been show by studies that failing to train new hires results in 52% more accidents. When no outside source for training is available, it results in 59% more accidents. The failure to adequately train supervisors results in 62% more accidents. According to a Nebraska Safety Council 1981 study comparing companies that did training with those that did not, these figures further support why training should be an integral part of any OSH imitative.

Why do employers need a training program for safety and health, and what does it entail? Before beginning an education and training program, make sure that training is needed by performing either a needs assessment or a cost–benefit analysis to determine if training will make a difference considering the dollars invested.

The reasons why the training should be an organized approach are not only the investment in dollars or cost, but also, if specific learning outcomes do not exist, then training would be a waste of resources, including time and loss of production. Also, learning objectives that mirror learning and performance expectations regarding OSH are an integral part any safety and health program.

OSH training can be used to address specific hazards within the employer's workplaces and structure the presentation of training materials that will best accomplish the goals of the employer and the training needs of the workforce. This may include classroom training, use of experts, use of qualified supervisors or employees, job instruction training (JIT), or on-the-job training (OJT).

Why Train New Employees?

The standardization of learning for everyone is the why formal training is desirable. This is especially important for newly hired employees. History has shown that individuals new to the workplace suffer more injuries and deaths than experienced workers. These usually occur within days or the first month on the job. It is imperative that new hires receive initial training regarding safety and health practices at the work site. New-hire training sets the stage or lays the foundation of the employer's commitment to OSH. New-hire training should include the following:

- Accident reporting procedures
- Basic hazard identification and reporting

- Chemical safety
- Company's basic philosophy on safety and health
- Company's safety and health rules
- Confined space entry
- Electrical safety
- Emergency response procedures (fire, spills, etc.)
- Eyewash and shower locations
- Fall protection
- Fire prevention and protection
- First aid/cardiopulmonary resuscitation (CPR)
- Hand tool safety
- Hazard communications
- Housekeeping
- Injury reporting procedures
- Ladder safety
- Lockout/tagout procedures
- Machine guarding
- Machine safety
- Material handling
- Medical facility location
- Mobile equipment
- Personal responsibility for safety
- Rules regarding dress code, conduct, and expectations
- Reporting procedures for unsafe acts/conditions
- Use of personal protective equipment (PPE)

Why Train Supervisors?

It is important that supervisors also receive useful training since the employer's OSH message and philosophy flow directly from the supervisor to the employees. The supervisor is the key person in the accomplishment of safety and health in the workplace. The supervisor will often be overlooked when it comes to safety and health training. Thus, the supervisor is frequently ill equipped to be the lead person for the company's safety and health initiative. A list of suggested training topics for supervisors includes the following:

- Accident causes and basic remedies
- Building attitudes favorable to safety
- Communicating safe work practices
- Cost of accidents and their effect on production

- Determining accident causes
- First-aid training
- Giving job instruction
- Job instruction for safety
- Knowledge of federal and state laws
- Making the workplace safe
- Mechanical safeguarding
- Motivating safe work practices
- Number and kinds of accidents
- Organization and operation of a safety program
- Safe handling of materials
- Supervising safe performance on the job
- Supervisor's place in accident prevention
- The investigation and methods of reporting incidents to the company and government agencies

All managers should receive formal training regarding the safety and health program management and responsibilities similar to requirements for supervisors.

Why Train Employees?

The training of all workers in safety and health has been demonstrated to reduce costs and increase the bottom line. Use the following guidelines to insure that employees are trained and safety and health training has been upgraded:

- Any new hazards or subjects of importance
- Basic skills training
- Explanation of policies and responsibilities
- Federal and state laws
- First-aid training
- Importance of first-aid treatment
- Methods of reporting accidents
- New safe operating procedures
- New safety rules and practices
- New skill training for new equipment, etc.
- Technical instruction and job descriptions

It is always a good idea for the employer, as well as the worker, to keep records and documents of training. These records may be used by a compliance inspector during an inspection, after an accident resulting in injury or illness, as a proof of good intentions by the employer to comply with training requirements for workers, including new workers and those assigned new tasks.

Why Communications?

Communication is the key to OSH. The training message of accident prevention must constantly be reinforced. Use constant reminders in the form of message boards, fliers, newsletters, paycheck inserts, posters, the spoken word, safety talks, meetings, and face-to-face encounters. Make the message consistent with the policies and practices of the company in order for it to be believable and credible. It is important to communicate safety and health goals and provide feedback progress toward accomplishing those goals. Most workers want their information in short, easily digested units.

Face-to-face interactions personalize the communications, provide information immediately (no delays), and allow for two-way communications, which improves the accuracy of the message. Finally, these interactions provide for performance and real-time feedback and reinforcement regarding safety and health.

Why Is Training a Key Element?

Training is one of the most important elements of any safety and health program. Each training item should describe methods for introducing and communicating new ideas into the workplace, reinforcing existing ideas and procedures, and implementing the safety and health program into action. The training needs may include manager and supervisor training, worker task training, employee updates, OJT, hands-on training, and new-worker orientation. The content of new-worker or new-site training should include at least the following topics:

- Company safety and health program and policy
- Employee and supervisory responsibilities
- Hazard communication training
- Emergency and evacuation procedures
- Location of first-aid stations and fire extinguishers and emergency telephone numbers
- Site-specific hazards
- Procedures for reporting injuries
- Use of PPE
- Hazard identification and reporting procedures
- Review of each safety and health rule applicable to the job
- Site tour or map where appropriate

It is a good idea to have follow-up for all training, which may include working with a more experienced worker, supervisor coaching, job observations, and reinforced good/ safe work practices.

Why OSHA Training?

Many of the OSHA regulations have specific requirements on training for fall protection, hazard communication, hazardous waste, asbestos and lead abatement, scaffolding, etc. It seems relatively safe to say that OSHA expects workers to have training on general safety and health provisions and hazard recognition, as well as task-specific training. Training workers regarding safety and health is one of the most effective accident prevention techniques.

Many standards promulgated by OSHA explicitly require the employer to train employees in the safety and health aspects of their jobs. Other OSHA standards make it the employer's responsibility to limit certain job assignments to employees who are "certified," "competent," or "qualified"—meaning that they have had special previous training, in or out of the workplace. The term *designated* personnel means selected or assigned by the employer or the employer's representative as being qualified to perform specific duties. These requirements reflect OSHA's belief that training is an essential part of every employer's safety and health program for protecting workers from injuries and illnesses.

As an example of the trend in OSHA's safety and health training requirements, the Process Safety Management of Highly Hazardous Chemicals standard (29 CFR 1910.119) contains several training requirements. This standard was promulgated under the requirements of the Clean Air Act Amendments of 1990. The Process Safety Management standard requires the employer to evaluate or verify that employees comprehend the training given to them. This means that the training to be given must have established goals and objectives regarding what are to be accomplished. Subsequent to the training, an evaluation would be conducted to verify that the employees understood the subjects presented or acquired the desired skills or knowledge. If the established goals and objectives of the training program were not achieved as expected, the employer then would revise the training program to make it more effective or conduct further sequent refresher training, or some combination of these. The requirements of the Process Safety Management standard follow the concepts embodied in the OSHA training guideline.

The length and complexity of OSHA standards may make it difficult to find all the regulations for training. To help employers, safety and health professionals, training directors, and others with a need to know, OSHA's training-related requirements are found in *Training Requirements in OSHA Standards and Training Guidelines (OSHA 2254)*. For more detail, consult the Code of Federal Regulations.

It is usually a good idea for the employer to keep a record of all safety and health training. Records can provide evidence of the employer's good faith and compliance with OSHA standards. Documentation can also supply an answer to one of the first questions an accident investigator will ask: "Was the injured employee trained to do the job?"

Training in the proper performance of a job is time and money well spent, and the employer might regard it as an investment rather than an expense. An effective program of safety and health training for workers can result in fewer injuries and illnesses, better morale, and lower insurance premiums, among other benefits.

Why a Legal Basis for Training?

The adequacy of employee training may also become an issue in contested cases where the affirmative defense of unpreventable employee misconduct is raised. Under case law well established in the Occupational Safety and Health Review Commission and the courts, an employer may successfully defend against an otherwise valid citation by demonstrating that all feasible steps were taken to avoid the occurrence of the hazard and that actions of the employee involved in the violation were a departure from a uniformly and effectively enforced work rule of which the employee had either actual or constructive knowledge.

In either type of case, the adequacy of the training given to employees in connection with a specific hazard is a factual matter, which can be decided only by considering all the facts and circumstances surrounding the alleged violation. The guidelines are not intended, and cannot be used, as evidence of the appropriate level of training in litigation involving either the training requirements of OSHA standards or affirmative defenses based upon employer training programs.

In an attempt to assist employers with their occupational safety and health training activities, OSHA has developed a set of training guidelines in the form of a model. This model is designed to help employers develop instructional programs as part of their total education and training effort. The model addresses the questions of who should be trained, on what topics, and for what purposes. It also helps employers determine how effective the program has been and enables them to identify employees who are in greatest need of education and training. The model is general enough to be used in any area of OSH training and allows employers to determine for themselves the content and format of training. Use of this model in training activities is just one of many ways that employers can comply with the OSHA standards that relate to training and enhance the safety and health of their employees. The training model can be found in *Training Requirements in OSHA Standards and Training Guidelines (OSHA 2254).*

Summary

Safety and health training is not the answer to accident prevention. Training cannot be the sum-total answer to the accidents/incidents that are occurring within the workplace. In fact, training is only applicable when a worker has not been trained previously, a worker is new to a job or task, or safe job skills need to be upgraded.

Companies may offer all types of programs and use many of the recognized accident prevention techniques, but without workers who are trained in their jobs and have safe work practices, efforts to reduce and prevent accidents and injuries will result in marginal success. If a worker has not been trained to do his/her job in a productive and safe manner, a very real problem exists.

Do not assume that a worker knows how to do his/her job and will do it safely unless he/she has been trained to do so. Even with training, some may resist safety procedures, and this presents a deportment or behavioral problem and not a training issue.

It is always a good practice to train newly hired workers or experienced workers who have been transferred to a new job. It is also important that any time a new procedure, new equipment, or extensive changes in job activities occur, workers receive training. Well-trained workers are more productive, more efficient, and safer.

The safety and health training that is needed should include not only workers but also supervisors and management. Without training for managers and supervisors, it cannot be expected or assumed that they are cognizant of the safety and health practices of your company. Without this knowledge, they will not know safe from unsafe, how to implement the loss control program, or even how to reinforce, recognize, or enforce safe work procedures unless given proper OSH training.

Training for the sake of documentation is a waste of time and money. Training should be purposeful and goal or objective driven. An organized approach to on-the-job safety and health will yield the proper ammunition to determine the company's real training needs. These needs should be based on accidents/incidents, identified hazards, hazard/accident prevention initiatives, and input from your workforce. Tailor training to meet the company's needs and that of the workers.

Look for results from the training. Evaluate those results by looking at the reduced number of accidents/incidents, improved production, and good safety practices performed by the workforce. Evaluate the results by using job safety observations and safety and health audits, as well as statistical information on the numbers of accidents and incidents.

Do not construe these previous statements to suggest that safety and health training does not have an important function as part of an accident prevention program. Without a safety and health training program, a vital element of workplace safety and health is missing.

Further Readings

Reese, C.D. *Accident/Incident Prevention Techniques (Second Edition)*. Boca Raton: CRC Press/Taylor & Francis, 2011.

Reese, C.D. *Occupational Health and Safety Management (Third Edition)*. Boca Raton, FL: CRC Press, 2016.

Reese, C.D. and J.V. Eidson. *Handbook of OSHA Construction Safety & Health (Second Edition)*. Boca Raton, FL: CRC Press/Taylor & Francis, 2006.

United States Department of Labor. *Training Requirements in OSHA Standards and Training Guidelines (OSHA 2254)*. Washington, DC: US Department of Labor, 1998.

United States Department of Labor, Occupational Safety and Health Administration, Office of Training and Education. *OSHA Voluntary Compliance Outreach Program: Instructors Reference Manual*. Des Plaines IL: US Department of Labor, 1993.

Section C

Administration

There need to be programs and elements that give guidance on the management of occupational safety and health (OSH). These include principles and philosophies under which it functions as well as specifics that iterate the intent and how it fits into a practical approach of accomplishing its goals and ideals.

The contents of this section are as follows:

Chapter 11—Safety and Health Budget

Chapter 12—Statistics and Tracking

Chapter 13—Safety and Health Ethics

Chapter 14—Employee Involvement

Chapter 15—Joint Labor/Management Safety and Health Committees

Chapter 16—Workplace Inspections

11

Safety and Health Budget

Occupational safety and health (OSH) is an element that should be managed the same as any other company function. As a unique and separate management undertaking, it must have a separate and distinct budget with dollars allocated for the various responsibilities, mandates, and requirements placed on the OSH initiative. Some of the reasons why a safety and health budget is needed are as follows:

- To prevent negative impact from production pressures
- To place monetary value on safety and health
- To make safety and health a part of managing the company
- To provide financial emphasis on safety and health needs
- To make safety and health an integral part of the company's financial planning
- To show management's commitment

Without an extensive history of cost related to OSH, the development of a budget for safety and health can be a rather inexact undertaking. Without such a history of past budgets or spending on OSH, budgeting becomes somewhat of a guesstimate. This is not to suggest that a reasonable and logical budget cannot be formulated. However, the budget will require much more effort, research, and justification.

Over the years, there have been few seminars or classes regarding budgeting for OSH, and the only book that gives the topic a better-than-adequate coverage is *Safety and Health Management Planning* by James P. Kohn and Theodore S. Ferry. It seems safe to say that most safety and health professionals have had little or no training or experience related to budgeting for OSH unless they have had hands-on experience from having to develop budgets. This is a function of doing while learning and probably results in some painful and time-consuming lessons.

Budgets are usually considered to be a road map or planning document for the completion of the assigned tasks and responsibilities based on the resources allocated by the company and are never all that they should be or what a safety and health professional would want them to be.

Budget items should include the following:

- Personnel cost
- Equipment (both expendable and nonexpendable)
- Travel
- Administrative cost
- Compliance cost
- Contracts (e.g., hazardous waste disposal)
- Facilities
- Liability insurance

- Budgeting for hiring new personnel
- Budgeting for long-term or multiyear projects
- Allowance for unforeseen emergencies
- Cost for Occupational Safety and Health Administration (OSHA) citations and violations

While it would be great to be able to predict the future and plan for events and items as though they would certainly happen, it is unlikely that the use of the budget as a precise document to follow will occur. Many factors are not in the control of the person responsible for OSH. Thus, the developers of an OSH budget must hitch their proverbial wagon to as many real-life safety and health issues as possible. This means that the best approach is to tie as much of the expenditures to compliance with regulatory requirements as humanly possible. Another approach is to show the use of dollars as intervention and prevention of potential cost (a cost avoidance strategy). A budget that is developed identifies specific items that are to be completed for a specific cost with proper justification and provides the resources to complete the agreed-upon safety and health task. A budget should be broken down into several identifiable categories, such as the following:

- Workplace health issues
- Workplace safety issues
- Safety and health management issues
- Environmental safety issues
- Product safety issues

Health Budgeting

Health budgeting must address the existing health factors and the required controls and need for sampling or screening. This may require that an industrial hygienist be hired full time, part time, or on contract. The cost of purchasing sampling instruments and maintaining their calibration as well as the expense of having samples analyzed by a certified laboratory must be factored into the budget.

The best use of dollar resources will determine the best approach to the effective use of the money available. Programs that may require sampling are hazardous chemicals, radiation, and noise. If new chemicals are to be put to use, this may require more sampling than previously. If OSHA citations have been issued, then more sampling may be necessary to maintain compliance. The presence of hazardous waste may require training in spill containment and remediation as well as contracts for disposal and spill cleanup or remediation. The health issues involved in the company's operation may result in special types of personal protective equipment being needed and purchased. Training will be needed on the use of the equipment, which is a cost factor in loss of production and time and must be accounted for.

Safety Budgeting

Safety budgeting is a function of energy since the sources of accident or incidents are usually the result of energy being released. The budget should include the cost of placing safeguards on machines and equipment, protection from exposure to electricity, protection and training of equipment operators with controls or personal protective equipment, the damage to equipment, implementation of controls, fire prevention and firefighting needs, the actual cost of injuries as well as disruption of production, and recertification of equipment (e.g., cranes). Any time a safety incident can be prevented, a cost saving is accomplished. The safety budget should emphasize the cost of safety prevention activities versus the cost of occurrences of safety-related incidents.

Management Budgeting

Safety and health management is more of an administrative function even though action items and their completion are an integral part of the safety and health initiative. The functions of time and cost for staff and workers involved in safety meetings, safety talks, and participatory programs are an indicator of the company's commitment to safety and health.

Cost and time to conduct audits and inspections to identify hazards and make recommendations for intervention and controls are management functions. The emphasis on time may seem irrelevant, but time is money or cost. Audits may result in the need for new equipment, revamping of processes and procedures, reengineering of equipment, revising of programs, training or retraining, or changes in safe operating procedures.

As a safety and health professional in charge, you must make sure that there is the proper staff with the right skill levels to perform the action elements contained within the budget that has been developed. If the company does not have the individuals with the skills needed, then these needed individuals can be contracted, or use temporary qualified staff. Include this cost in the budget.

When a company implements a new program that is needed to effectively reduce the risk of accidents/incidents, it results in an investment in resources again of time and dollars.

When an accident, incident, or fatality occurs, myriad time-consuming and costly factors accompany it—not just the cost of the investigation but the loss of production, equipment damage, loss of productive worker, litigation, OSHA enforcement activities, loss of supervisor's time; the list goes on, and costs mount.

There is always the cost of new-hire training, new-job or new-task training, new-process or new-program training, safety and health training for managers and supervisors, and refresher training for such areas as hazardous waste remediation, asbestos abatement, and lead abatement. At times, first-aid and cardiopulmonary resuscitation training or retraining is necessary. The components of managing safety and health must be planned for in the OSH budget.

Environmental Budgeting

In many companies, environmental safety is addressed as a combined effort with OSH, entitled environmental/occupational safety and health (EOSH). There is a link between the two since environmental issues can cause both safety and health hazards to workers, such as hazardous material exposure from spills or emergencies from the release of toxic gases. The development of a budget for environmental safety and health should follow a similar process as OSH because failure to comply with the US Environmental Protection Agency's regulations can be much more costly than failure to comply with OSHA's regulations with regard to appreciable higher fine assessment for violations.

Product Safety Budgeting

Product safety is not an item that should be the responsibility of the OSH department. It should be the responsibility of the product development, quality control, or consumer safety personnel within the company since the applicable standards are found in a regulatory arena different from either the environment or OSH. This not to suggest that EOSH professionals should not lend their expertise to other departments, when they can contribute. Product safety should not be part of the EOSH budget, as product safety does not deal with worker safety and health.

Compliance Factors

The OSHA database regarding the average cost of a citation can be used as a lever to get management's attention to the investment in safety and health requested in the budget. It is a real eye-opener for management when safety and health professionals can say that the cost of compliance is $500 for blood-borne pathogens and the average violation produces a fine of $1000 per citation. In addition, the OSH professionals can research the most frequent violation for their particular industry sector. In many cases, the most frequent violation is the requirement for a program and training under the hazard communication standard.

Using compliance is by far one of the best ways to justify expenditures. Some of the regulations that require programs for compliance are as follows:

- Personal protective equipment
- Respiratory protection program
- Hazmat program
- Employee emergency and fire prevention plans
- Blood-borne pathogen program
- Medical and first aid
- Ventilation systems

Some of the safety and health regulations require medical examinations to be in compliance, such as the following:

- Hearing examinations
- Hazardous materials and waste
- Respiratory clearance examinations

An OSHA violation or citation strikes fear into most employers and is an excellent lever to justify budget requests. Other such levers include the following:

- Cost of accidents and incidents
- Medical costs
- Workers' compensation costs
- Real dollar savings
- Loss of potential from not performing the action item

Written Budget

When it is time to put a budget in writing, make sure to make good judgment decisions, which means to use all the resources available, such as historical-based data, and have the resources needed to fulfill the program requested within the budget. Make sure that the company is getting the "best bang for the buck." Is it really worth doing a certain task? Is there a gain in dollars or other benefits? Could it have long-range cost benefits? Is there going to be a return on the money and resources invested? Is the spending going to accomplish your goals? Answers to these questions will help you prioritize the items that should receive emphasis in the budget.

Whether developing an element or task budget, a variable budget, or a percent-of-allocation-by-departments budget, it is important to identify operating costs and potential benefits such as direct benefits (reduced labor cost, lower rate of accidents, reduced insurance cost, or productivity gains). Indirect benefits should also be considered in the light of quality improvements—reduced scrap; less rework; reduced product liability, exposure, or product recall expenses; improved corporate image; or increased market share. At times, indirect benefits have improved employee morale, which reduces absenteeism and increases turnover, teamwork, and ownership. The potential reduction in the numbers of compliance penalties can be a benefit.

Controlling Cost

The safety and health professional who develops a responsible budget that looks for ways to control cost and maintain or improve the company's bottom line will be viewed as a part of the team and not just a necessary evil.

The safety and health professional can look at ways to control cost by sharing resources, using his/her available resources to the fullest, working on projects with other departments, lending expertise when possible, using mail order to cut cost, using cost-bid service to obtain more for dollars spent, using in-house engineering expertise when possible, volunteering use of staff expertise when possible, using consultants or contractors for temporary or short-term projects for cost saving, performing as much as possible with own staff, being productive, and taking low-cost action by being a good shopper.

Summary

There is always a degree of inaccuracy in any budget. Prevent cost overruns if possible since overruns almost always have detrimental effects. Make sure to use the most qualified person to develop the OSH budget. Remember that there will always be budget expenditure and issues beyond the OSH manager's control since actual cost may not be the same as planned cost. Link budget elements to accomplishments, milestones, needs, and compliance. Provide as much substance as possible in the budget. Quantified justifications have more substance than qualified opinions. Numbers are more understandable to senior management than reducing severity rates. Use all the experience and tools to develop a workable and reasonable budget.

Further Readings

Allison, W.W. *Profitable Risk Control*. 1986. Des Plaines, IL: American Society of Safety Engineers, 1986.

Brigham, E.F. *Fundamentals of Financial Management*. Chicago, IL: Dryden, 1989.

Kohn, J.P. and T.S. Ferry. *Safety and Health Management Planning*. Rockville, MD: Government Institutes, 1999.

Reese, C.D. *Accident/Incident Prevention Techniques (Second Edition)*. Boca Raton, FL: CRC Press, 2012.

Reese, C.D. *Occupational Health and Safety Management (Third Edition)*. Boca Raton, FL: CRC Press, 2016.

Waxman, K.T. *A Practical Guide to Finance & Budget: Skills for Nurse Managers*. Marblehead, MD: HCPro Inc., 2004.

12

Statistics and Tracking

Until a business or industry tracks accidents, their related costs, and the resulting injuries and illnesses, they have little baseline information upon which to lobby for better accident prevention, which can result in a multitude of benefits. Some of these benefits are as follows:

- Better Occupational Safety and Health Administration (OSHA) compliance
- Economic improvement
- Humanitarian benefits
- Less legal liability
- Better employer/employee relations
- More time for production

The benefits of tracking a company's workplace safety and health accidents/incidents are that it provides the data to assess the effectiveness of any accident prevention initiative and allows the company to determine the cost benefits of the safety and health dollars that are being spent by the company. Thus, record keeping and tracking make good business sense.

The analysis of industry-related hazards and the accidents/incidents that they cause are an important step in the overall process of reducing occupation-related injuries, illnesses, and deaths. Only after a systematic look at the hazards and accidents can a business or industry hope to integrate the accident prevention techniques and tools that can have an impact upon a company's safety and health initiative.

Analyzing Incident Data

The importance of monitoring and tracking the overall safety and health efforts of companies cannot be overstated. If companies do not keep records on all aspects of its accident prevention effort, they will not have the information needed to obtain continuous funding for accident prevention or justification of its performance value. Those companies will not be able to explain where budget dollars were expended, nor will they have a way to assess the loss-control progress within the company. The data for comparing a company's performance against trends and tendencies of occupational injuries and illnesses occurring in other companies or nationally come from the records being kept by individual companies.

Tracking should not just be a count of recordable incidents. Companies should track near misses, equipment damage, and first-aid events.

Tracking means that company officials must pay close attention to workers' compensation cases as well as their associated costs. The longer an individual is off the job, the more the cost, and the greater the likelihood that the worker will not return to work.

OSHA Record Keeping

As part of the tracking process, a company must comply with OSHA's record-keeping requirements for occupational injuries and illnesses. Any occupational illness that has resulted in an abnormal condition or disorder caused by exposure to environmental factors, which may be acute or chronic due to inhalation, absorption, ingestion, or direct contact with toxic substances or harmful agents, and any repetitive motion injury, is to be classified as an illness. All illnesses are recordable, regardless of severity. Injuries are recordable under the following conditions:

1. An on-the-job death occurs (regardless of length of time between injury and death).
2. One or more lost workdays occur.
3. Restriction of work or motion transpires.
4. Loss of consciousness occurs.
5. The worker is transferred to another job.
6. The worker receives medical treatment beyond first aid.

Company Records

Many companies conduct accident investigations and keep accident records and other data on the company's safety and health initiatives. If a company has a sufficient number of accidents/incidents and enough detail in their occupational injury/illness investigation data, the company can begin to examine trends or emerging issues relevant to their safety and health intervention/prevention effort. The analysis of these data can be used to evaluate the effectiveness of safety and health at various workplaces and jobsites, or for groups of workers. The safety and health data can be used by a company to compare to other companies that perform similar work, employ a comparable workforce, or compete in the same kind and size of market on a state, regional, national, or international basis.

By analyzing the company's accidents/incidents, companies are in a better position to compare apples to apples rather than apples to oranges. Companies not only will be able to identify the types of injuries, types of accidents, and types of causes, but they will also be able to intervene and provide recommendations for preventing these accidents/incidents in the future. Companies will be able to say with confidence that "I do" or "I do not" have a safety and health problem. If a company finds that it has a problem, the analysis and data will be essential, especially if it is trying to elicit advice on how to address their safety and health needs.

Gathering and analyzing accident/incident data is not the company's entire safety and health program but a single element. Data provide feedback and evaluative information as companies proceed toward accomplishing their safety and health goals; thus, data contribute an important component in the analysis process.

Important Ancillary Data Needed for More Complete Analysis

Careful and complete analysis of the data collected following an incident is critical to the accurate determination of an accident's causal factors and an important part of preventing a reoccurrence.

The results of comprehensive analyses provide the basis for corrective and preventive measures.

The analysis portion of the accident investigation is not a single, distinct part of the investigation. Instead, it is the central part of the process that includes collecting facts and determining causal factors. Well-chosen and carefully performed analysis is important since it provides results that can be used by a company to improve its safety and health performance.

When collecting data after an incident, the volume seems to be ever increasing as the incident is evaluated. The accuracy and relevance of the data is critical to an effective analysis process. Some of the data items that should be acquired are as follows:

- Age
- Sex
- Race
- Date
- Time of the incident
- Day of the week
- The shift
- Occupation of victim
- Task being performed at time of incident
- Type of accident
- Nature of injury or illness—names the principal physical characteristic of a disabling condition, such as sprain/strain, cut/laceration, or carpal tunnel syndrome
- Part of body affected—directly linked to the nature of injury or illness cited, such as back, finger, or eye
- Source of injury or illness—the object, substance, exposure, or bodily motion that directly produced or inflicted the disabling condition cited. Examples include lifting a heavy box; exposure to a toxic substance, fire, or flame; and bodily motion of an injured or ill worker

Safety and Health Statistics Data

The Bureau of Labor Statistics' (BLS's) *Occupational Injury and Illness Classification System Manual* provides a much more detailed breakdown of the type of data that is most useful in analysis of occupational incidents. The four major areas of data that the BLS believes are important to collect during review or investigation of OSH incidents are nature, part of body, source, and event or exposure.

Each of these is further broken down in greater detail in BLS's *Occupational Injury and Illness Classification System Manual*. The BLS also has a manual that details occupations.

The principal challenge in collecting data is distinguishing between accurate and erroneous information in order to focus on areas that will lead to identifying the accident's causal factors. These causal factors are keys to accident/incident prevention. This can be accomplished by the following:

- Understanding the activity that was being performed at the time of the accident
- Personally conducting a walk-through of the accident scene
- Challenging "facts" that are inconsistent with other evidence (e.g., physical)
- Corroborating facts through interviews
- Testing or inspecting pertinent components to determine failure modes and physical evidence
- Reviewing policies, procedures, and work records to determine the level of compliance or implementation.

Careful and complete analysis of the data collected following an accident is critical to the accurate determination of an accident's causal factors. The results of comprehensive analyses provide the basis for corrective and preventive measures.

Statistical Analysis for Comparisons

Companies and federal officials frequently utilize the following statistical pieces of information, which are designed to allow the company to compare its safety and health performance with others: the incident rate, illness rate, lost workday case rate or severity rate, and restricted workday case rate. These rates, answer the following questions: "How often or frequently are accidents occurring?" "How bad are the injuries/illnesses that are occurring?" The number of times that occupational injuries/illnesses happen is the determinant for the incident rate, while the number of days away from work (lost-time workdays) or restricted workdays is the prime indicator of the severity rate. Both of these rates provide unique information regarding your safety and health effort.

Workers' Compensation

Each employer is expected to provide protection for his/her workers who become ill or are injured by something within the workplace. The premiums that the employer pays are to provide medical treatment and supplemental income when a worker is unable to return to work immediately. This supplemental income is usually two-thirds of the worker's wages and is not taxed. Once a worker files a claim for workers' compensation, the employee is not usually allowed to sue his/her employer.

Thus, workers' compensation premiums paid by the employer act like a protection or insurance policy against liability when a worker suffers injury or illness due to a hazard within the workplace.

The employer's premium is based upon the type of industry as well as the injury and illness history of the employer's workplace. The more injuries and illnesses employers experience, the higher the premium paid by the employer. Also, the more days away from work experienced by the workers, the more the premium increases.

Thus, as a cost factor, the reduction of occupational injuries and illnesses is important to cost containment. Any responsible employer understands the relationship between the bottom-line profit and the number of accidents.

Employers must set up a tracking system, which monitors each workers' compensation claim filed by their workers. They also need to work diligently to return workers back to the workforce as soon as medically feasible. As part of the healing process, the employer needs to make sure that the workers are receiving their medication and going for medical treatment. The longer a worker is away from the workplace, the more unlikely that he/she will return to the workforce. It has been documented that workers who are off work for 6 months have a 50% chance of returning to work. For those off a year, the probability is 25%, and after 2 years or longer, the probability of returning to work is 0%.

Each workers' compensation claim is either directly or indirectly a drain upon the profit margin of the company. This is the primary business reason to track workers' compensation claims. Many employers who have a less-than-stellar experience modification rate (EMR) may have difficulty in procuring work, or bidding on work, when perspective buyers assess their safety and health performance.

The prevention of occupational injuries and illnesses will definitely reduce the cost of workers' compensation premiums. It does not matter whether the company pays into the state workers' compensation system or is self-insured. What motivates most employers to reduce workplace injuries and illnesses is usually dollars. Some studies show that other benefits result from a decrease in the number of injuries and illnesses. For example, employers can expect to increase attendance, morale, and productivity. These are just a few of the side effects of reduced workers' compensation cost.

Summary

Statistical information lends credibility to the OSH effort of a company. These numbers provide substance to recommendations, suggestions, progress, accomplishments, and direction for a company's OSH initiative.

Further Readings

Reese, C.D. *Accident/Incident Prevention Techniques (Second Edition)*. Boca Raton, FL: CRC Press, 2012.

Reese, C.D. *Occupational Health and Safety Management (Third Edition)*. Boca Raton, FL: CRC Press, 2016.

Reese, C.D. and J.V. Eidson. *Handbook of OSHA Construction Safety & Health (Second Edition)*. Boca Raton, FL: CRC/Lewis Publishers, 2006.

United States Department of Labor, Occupational Safety and Health Administration, Office of Training and Education. *OSHA Voluntary Compliance Outreach Program: Instructors Reference Manual*. Des Plaines, IL: US Department of Labor, 1993.

13

Safety and Health Ethics

Certainly, it seems that ethics is not a topic to discuss when occupational safety and health (OSH) surfaces as a subject. But, this is why it is part of the big picture in giving legs to the relevance of OSH. The answer as to why ethics is a key principle to OSH is as follows:

- Safety and health must be a set of ethical and moral values.
- Failure to set standards and values degrades safety and health credibility.
- Loss of safety and health credibility will be hard to recover from.
- OSH is held to a higher standard.
- Commitment to safety and health is demonstrated by actions.
- Safety and health values or attitudes are exhibited in behaviors toward safety and health.
- What is done in the name of OSH speaks louder than what is said.
- Much can be lost when no true value is placed on safety and health.
- A simple failure to ethically represent OSH does much to decrease its stature.

There is an old saying that goes like this: "Credibility is a lot like virginity," in that once it is lost, it is gone. Credibility, once lost, is nearly impossible to regain, and in this statement, the same holds true for the ethics of safety and health conduct. If a company or the person responsible for safety and health (i.e., safety manager) breaches the ethics regarding safety and health of workers, the words "ethical behavior" may just as well be a foreign language. The trust or credibility that existed is gone and will take a great deal of effort to recover.

Safety and health is viewed as a profession that has at its core responsibility for the health and welfare of people. This puts the safety and health professional on a similar pedestal as a medical professional. Safety and health professionals are expected to function under a code of ethics and exhibit behavior that is compatible with that code.

Ethics

The US military provides a glimpse of what is meant by ethics. Ethics are standards by which one should act based on values. Values are core beliefs such as duty, honor, and integrity that motivate attitudes and actions. Not all values are ethical values (integrity is; happiness is not). Ethical values relate to and thus take precedence over nonethical values when making ethical decisions.

Primary ethical values include the following:

- Honesty. Being truthful, straightforward, and candid are aspects of honesty. Truthfulness is required. Deceptions are usually easily uncovered. Lies erode credibility and undermine public confidence. Untruths told for seemingly altruistic reasons (to prevent hurt feelings, to promote goodwill, etc.) are nonetheless resented by the recipients. Straightforwardness adds frankness to truthfulness and is usually necessary to promote public confidence and to ensure effective, efficient conduct of operations. Truths presented in such a way as to lead recipients to confusion, misinterpretation, or inaccurate conclusions are not productive. Such indirect deceptions can promote ill will and erode openness, especially when there is an expectation of frankness. Candor is the forthright offering of unrequested information. It is necessary according to the gravity of the situation and the nature of the relationships. Candor is required when a reasonable person would feel betrayed if the information were withheld. In some circumstances, silence is dishonest; yet in other circumstances, disclosing information would be wrong and perhaps unlawful.

- Integrity. Being faithful to one's convictions is part of integrity. Following principles, acting with honor, maintaining independent judgment, and performing duties with impartiality help to maintain integrity and avoid conflicts of interest and hypocrisy.

- Loyalty. Fidelity, faithfulness, allegiance, and devotion are all synonyms for loyalty. Loyalty is the bond that holds nations and companies together and the hedge against dissension and conflict. Loyalty is not blind obedience or unquestioning acceptance of the status quo. Loyalty requires careful balancing among various interests, values, and institutions in the interest of harmony and cohesion.

- Accountability. Professionals are required to accept responsibility for their decisions and the resulting consequences. This includes avoiding even the appearance of impropriety. Accountability promotes careful, well-thought-out decision making and limits thoughtless action.

- Fairness. Open-mindedness and impartiality are important aspects of fairness. Safety and health professionals must be committed to justice in the performance of their official duties. Decisions must not be arbitrary, capricious, or biased. Individuals must be treated equally and with tolerance.

- Caring. Compassion is an essential element of when individuals are being served. Courtesy and kindness, both to those we serve and to those we work with, help to ensure individuals are not treated solely as a means to an end. Caring for others is the counterbalance against the temptation to pursue the expected outcomes at any cost.

- Respect. To treat people with dignity, to honor privacy, and to allow self-determination are critical in a corporate environment of diverse people. Lack of respect leads to a breakdown of loyalty and honesty within a company and brings chaos to functionality.

- Promise keeping. No company can function for long if its commitments are not kept. Safety and health professionals are obligated to keep their promises in order to promote trust and cooperation. Because of the importance of promise keeping, safety and health professionals must only make commitments within their authority.

Occupational Safety and Health Ethics

The major factors affecting ethics related to safety and health are character and culture. If safety and health is not a value within the company culture, then a safety and health professional with all the character in the world cannot be effective. Safety and health must have support within the company; then, a safety and health professional can do what is right. This means that the professional must ask himself/herself the following trite questions:

- Would you be embarrassed if your friends and family found out what you did?
- What if this happened to you?
- Is it legal, and does it comply with regulatory requirements?
- Does it feel right?

If these question cannot be answered in the positive, then do what is right. This means that safety and health professionals must do the following:

- Really care about the welfare of everyone; it will show.
- Treat everyone and all aspects of safety and health equally.
- Treat everyone with respect and dignity.
- Take all concerns by employees seriously, and respond in a timely and professional manner.
- Try to avoid the gray areas concerning values and expectations.
- Set up acceptable standards for behavior regarding right and wrong safety and health issues.
- Lead by example.
- Do not accept unsafe acts or behavior.
- Continue to build trust and commitment.
- Reprisal is never an appropriate response.
- Hold themselves and others accountable for safety and health practices in an even-handed and fair manner.

Ethics are the moral principles and values that govern the actions and decisions of an individual or group. Operating a company according to the law does not necessarily mean that the practice is ethical. Again, the cornerstones of ethics are character and culture.

Further Readings

Reese, C.D. *Accident/Incident Prevention Techniques (Second Edition)*. Boca Raton, FL: CRC Press, 2012.
Reese, C.D. *Occupational Health and Safety Management (Third Edition)*. Boca Raton, FL: CRC Press, 2016.

14

Employee Involvement

In recent years, both the Occupational Safety and Health Administration (OSHA) and employers have come to realize that employee involvement is crucial in the conduct of an effective safety and health effort in the workplace. This is why true employee involvement is a mechanism that can assure commitment and buy-in by the workforce. Involvement can be a part of the office building environment or manufacturing operation to foster the kind of safety culture that goes a long way toward prevention of bad safety and health practices in the workplace.

Employee involvement is a very strong motivator for individuals in the workplace. It provides them with a say, some control, personal participation, and acknowledgement that they can contribute. All of these factors are very strong motivators, provide for linking to the safety and health goals of the company, and can result in a real commitment to safety and health.

Employee involvement provides the means through which employees develop and express their own commitment to safety and health, for both themselves and their fellow workers. It also will be of direct benefit to the employer through the following:

- Fewer employee injuries
- Decreased workers' compensation payments
- Decreased litigation costs relating to employee injury and illnesses
- Fewer OSHA noncompliance findings
- Improved employee morale
- Improved communication between management and employees
- Increased employee and management involvement in safety and health-related matters
- Increased productivity and profits
- Eliciting positive public relations

All of these factors can result in some very positive outcomes. A caution flag should be raised regarding the issue of employee involvement. With the growth and recognition of the value of employee involvement, there are an increasing number and variety of employee participation arrangements that may have to be addressed from a legal basis. It makes good sense to consult a labor relations advisor to ensure that the employee involvement program conforms to current legal requirements.

Why Should Employees Be Involved?

Because it's the right and smart thing to do. Here's why:

- Employees are the persons most in contact with potential safety and health hazards. They have a vested interest in effective protection programs.
- Group decisions have the advantage of the group's wider range of experience.
- Employees are more likely to support and use programs in which they have input.

Employees who are encouraged to offer their ideas and whose contributions are taken seriously are more satisfied and productive on the job.

Examples of employee participation include the following:

- Participating on joint labor management committees and other advisory or specific-purpose committees
- Conducting site inspections
- Analyzing routine hazards in each step of a job or process, and preparing safe work practices or controls to eliminate or reduce exposure
- Developing and revising the site safety and health rules
- Solving safety and health problems
- Training both current and newly hired employees
- Providing programs and presentations at safety and health meetings
- Conducting accident/incident investigations
- Reporting hazards
- Fixing hazards within their control
- Supporting their fellow workers by providing feedback on risks and assisting them in eliminating hazards
- Participating in accident/incident investigations
- Performing a preuse or change analysis for new equipment or processes in order to identify hazards up front before use

Employees want to feel that they have some control over their work lives and safety and health. Participation is one way to accomplish this. Everyone wants to belong and feel wanted. Employee involvement in safety and health can help accomplish this.

Why Does Management Have to Be Involved?

Management has to make the commitment to involve the workforce in safety and health at the work site. Since management is in control, it is their prerogative to decide the extent of involvement depending on what they can endure or the amount of control they can forgo. But there are very real positive outcomes that come from employees' involvement. It is a very real way to gain commitment to safety and health. Safety and health becomes a personal issue for your workforce.

When the door to involvement is opened, the benefits are good, but the Pandora's box that has opened may cause a decision to rethink this commitment. It may be found that many of the problems brought forth are not strictly what one might feel are safety and health issues, but to those in the workforce, they are what they perceive as legitimate subjects to be given consideration. Some of the areas that may come up are as follows:

- Workload (time pressures, work pressures, physical aspects of work)
- Job control (worker influence over his/her job)
- Social support (support of coworkers, supervisors, etc.)
- Job satisfaction (with the work and work environment)
- Monotonous work (bored with work)
- Job clarity (uncertainty of job expectations and ambiguity of worker's role)
- Psychological stress
- Organizational climate (e.g., high performance expectations)
- Organizational culture

Employee involvement will require a commitment by management as well as the feeling that such a commitment is genuine and not some sort of setup. It may require that employers provide training to employees who are expected to take on roles within the safety and health effort that they are not accustomed to. When everyone is committed, employee involvement can be a positive factor in managing the safety and health effort.

In the work environment, it is a good idea to have as many eyes as possible helping to ensure that safety and health policies and procedures are being adhered to. This is much easier to accomplish when an employer has the involvement and commitment of the workforce. This can be achieved by having all employees integrally involved, or possibly, it might be better to work through an organized approach such as a joint committee(s). It must be decided what matches the employer's management approach the best, but some sort of involvement is crucial to safety and health performance in the workplace environment.

Why Employee Involvement?

Any time something new is undertaken, such as involvement, there are expectations that accompany these new endeavors. Some of the expectations are as follows:

- Improved workplaces and work environment
- Improved working relationships
- Positive, cooperative approaches
- A compromise for mutual interests, versus self-serving interests
- A true team approach

- Sharing of information, thinking, and substantive decision making
- New/fresh ideas
- Increased participation and involvement

Why Are Employee Outcomes Important?

A study done by the Work in America Institute lists some of the outcomes that can be expected when a company has functional joint labor/management initiatives. According to this study, both labor and management stand to benefit from joint undertakings. Some of the benefits include the following:

- Economic gains: higher profits, less cost overruns, increased productivity, better quality, greater customer satisfaction, and fewer injuries and illnesses. Working together, workers and supervisors can solve problems, improve product quality, and streamline work processes.
- Improved worker capacities, which more effectively contribute to the improvement of the workplace.
- Human resource benefits.
- Innovations at the bargaining table.
- Committee member growth.
- Workplace democracy.
- Employment security.
- Positive perceptions.

Other outcomes that will, in all likelihood, arise from employee involvement are as follows:

- Shared responsibilities
- Increased individual involvement
- Company and labor being proactive with each other
- Better communications between company and labor
- Employee ownership of ideas, goals, activities, outcomes, and the company
- Union leadership and members being more challenged

Employee involvement can address specific workplace issues, such as the following:

- Monitor the safety and health programs
- Inspect the workplace to identify hazards
- Conduct and review accident investigations
- Recommend interventions and prevention initiatives
- Review injury and illness data for incident trends

- Act as a sounding board for workers who are expressing safety and health concerns
- Become involved in designing and planning for a safe and healthy workplace
- Make recommendations to the company regarding actions, solutions, and program needs for safety and health
- Participate and observe workplace exposure monitoring and medical surveillance programs
- Assure that training and education fully address safety and health issues facing the workplace

Why Are Goals Needed for Employee Involvement?

- To reduce accidents, through a cooperative effort, by eliminating as many workplace hazards as possible
- To reduce the number of safety and health-related complaints filed with regulatory agencies without infringing on the workers' federal and state rights
- To promote worker participation in all safety and health programs
- To promote training in the areas of recognition, avoidance, and the prevention of occupational hazards
- To establish another line of communication whereby the workers can voice their concerns regarding potential hazards and then receive feedback on the status or action being taken

Summary

Why is employee involvement critical to attaining better and more effective operations for a company?

- Increased commitment to achieving the organization's goals or mission.
- Improved productivity, safety and health, customer service, and product quality.
- Joint resolution of problems and issues facing the organization.
- Shared responsibility and accountability for results and outcomes.
- Better and more constructive relationship between labor and management.
- Enhanced employee morale and job satisfaction.
- Heightened communication and information sharing that brings all employees into the decision-making process. This helps them understand the mission, goals, and objectives of the organization and fosters employee support of the organization's undertakings.
- Increased job security and compensation.

Further Readings

Reese, C.D. *Accident/Incident Prevention Techniques (Second Edition)*. Boca Raton, FL: CRC Press, 2012.

Reese, C.D. *Occupational Health and Safety Management (Third Edition)*. Boca Raton, FL: CRC Press, 2016.

Reese, C.D. and J.V. Eidson. *Handbook of OSHA Construction Safety & Health (Second Edition)*. Boca Raton, FL: CRC/Lewis Publishers, 2006.

15

Joint Labor/Management Safety and Health Committees

A joint labor/management (L/M) committee is a formal committee that is organized to address specific issues such as safety and health or production processes. It is a committee with equal representation, which gives both parties an opportunity to talk directly to each other and educate each other concerning the problems faced by either group. In contrast, labor-only or management-only committees have self-serving goals with no consensus on solving problems. They are the only ones who have the authority or power to make changes. Thus, joint committees are aimed at gaining solutions, ensuring equal participation, and having some degree of authority or power.

Joint L/M committees have a different purpose from committees set up by either labor or management alone. When compared with other committees existing at the work site, L/M committees are different both in their goals and in methods of operation. In addition, because of the nature of their goals, they are also much more challenging since they require many different skills from all the participants.

Joint committees provide both parties with the opportunity and structure to discuss a wide range of issues challenging them. Neither partner of the L/M committee has enough information, commitment, or power to institute the changes that the joint committee eventually identifies as critical to the success of the business.

Thus, a key purpose of joint committees is to gather, review, analyze, and solve problems that are critical to the success of the business and are not appropriate for the collective bargaining process. Another purpose of these committees is the formation of a level playing field, which has as its ultimate purpose the success of the business. These committees can help build bridges of cooperation, which can lead to increased productivity, quality, efficiency, safety, health, and economic gains shared by all parties.

However, the purpose of joint L/M committees goes beyond quality and productivity. They are builders of true and honest relationships, which help to realize success through focusing on outcomes, using resources more efficiently, fostering real-world flexibility, supporting an information-sharing system, opening communications, and fostering a better working relationship.

In the past, L/M relationships were built on confrontation, distrust, acrimony, and the perception of loss or gain of control and power. The ultimate goal, to use an overused phrase, is to attain a win–win situation.

With all the downsizing, rightsizing, and reengineering going on within the workplace, individuals believe that they can help and have an impact on their continued employment and the survival of the company, if only asked and given the chance. To do this successfully, it is imperative that they have access to the information needed to solve issues facing them and their employer.

Committee Makeup

The joint L/M committee should be composed of at least as many employee members as employer members. Labor must have the sole right to appoint or select its own representatives, just as employers have the right to appoint theirs. Both parties should clearly understand that the members of this committee must not only be risk-takers but also be fully capable of making the critical decisions needed to make this process successful.

To cover all facets of the workplace, the labor organization may find it useful to have a broad spectrum of their membership represented. Labor should also allow for turnover in its membership and address this issue by identifying and involving adjunct and alternate members. By doing this, they will not compromise the committee's progress by introducing new members into the committee who are unfamiliar with the process and are untrained concerning the subject matter.

The chairperson must be elected by the committee, and this position should be a rotating position between labor and management. Each committee member should receive training on the joint committee process and receive other specific instructions that are deemed necessary, such as job-related safety and health training. The labor members should be paid for all committee duties, including attendance at meetings, inspections, training sessions, etc.

Record Keeping

Each participating party (labor and management) should keep its own notes of all meetings and inspections, as well as copies of agendas. This will ensure that agreements and disagreements, time schedules, actions to be taken, etc., are not lost, forgotten, neglected, or misinterpreted. Time has a way of encouraging each one of these things to happen. Good record keeping will also assist in keeping the direction and undertakings of the joint committee in focus. A formal set of minutes and reports on inspections should be maintained by the joint committee

Do's and *Don'ts* of L/M Committees

Do's:

- Always give an agenda to committee members in advance of a meeting; this allows everyone time for preparation.
- Cancel a meeting only for emergencies; hold meetings on schedule.
- Set timelines for solving problems.
- Focus on the issues involved.
- Do stay on schedule and stick to the starting and ending times in the agenda.

- Decide on a structured approach to recording and drafting minutes, as well as mechanisms for disseminating them.
- Keep the broader workforce informed of the activities of the committee.
- Keep issues on the agenda until they are resolved to everyone's satisfaction.
- Give worker representatives time to meet as a group and prepare for the meeting.
- Be on time for the meetings.
- Make sure that everyone understands the issues and problems to be discussed.

Don'ts:

- Tackle the most difficult problems first since some early successes will build a stronger foundation.
- Work on broadly defined issues, but deal with specific problems and concrete corrective actions.
- Allow the meeting to be a gripe session when problem solving is the result.
- Allow any issue to be viewed as trivial; each issue is important to someone.
- Let individual personalities interfere with either the meetings or the intent of the committee.
- Be a know-it-all and assume you know the answer; give everyone an opportunity to participate in solving the problem.
- Neglect to get all the facts before trying to solve an issue or problem.
- Prolong meetings.
- Delay conveying and communicating the solutions to problems and the outcomes or accomplishments that the committee has achieved.
- Expect miraculous successes or results immediately since many of the problems and issues did not occur overnight.

Expectations

Any time new endeavors are undertaken, such as joint L/M committees, there are expectations that accompany them. Some of the expectations are as follows:

- Improved workplaces and work environment
- Improved working relationships
- Positive, cooperative approaches
- A compromise of mutual interests, versus self-serving interests
- A true team approach
- Sharing of information, thinking, and substantive decision making
- New/fresh ideas
- Increased participation and involvement

Outcomes

A study done by the Work in America Institute lists some of the outcomes to be expected when there is a functional joint L/M committee. According to this study, both labor and management stand to benefit from joint undertakings. Some of the benefits include the following:

- Economic gains: higher profits, less cost overruns, increased productivity, better quality, greater customer satisfaction, and fewer injuries and illnesses.
- Working together, workers and supervisors can solve problems, improve product quality, and streamline work processes. Improved worker capacities, which more effectively contribute to the improvement of the workplace.
- Human resource benefits.
- Innovations at the bargaining table.
- Committee member growth.
- Workplace democracy.
- Employment security.
- Positive perceptions.

Other outcomes that will arise from joint L/M committees in all likelihood are as follows:

- Shared responsibilities
- Increased individual involvement
- Company and labor being proactive with each other
- Better communication between company and labor
- Employee ownership of ideas, goals, activities, outcomes, and the company
- Union leadership and members being more challenged

Joint L/M Occupational Safety and Health Committees

A joint L/M occupational safety and health (OSH) committee is a specialized application of the joint L/M committee and is an excellent format, which others can replicate. This type of committee is organized to address specific workplace issues such as the following:

- Monitor safety and health programs
- Inspect the workplace to identify hazards
- Conduct and review accident investigations
- Recommend interventions and prevention initiatives
- Review injury and illness data for incident trends
- Act as a sounding board for workers who express safety and health concerns

- Become involved in designing and planning for a safe and healthy workplace
- Make recommendations to the company regarding actions, solutions, and program needs for safety and health
- Participate in and observe workplace exposure monitoring and medical surveillance programs
- Assure that training and education fully address safety and health issues facing the workplace

The goals of the joint L/M OSH committees are as follows:

- To reduce accidents, through a cooperative effort, by eliminating as many workplace hazards as possible
- To reduce the number of safety- and health-related complaints filed with regulatory agencies without infringing on the workers' federal and state rights
- To promote worker participation in all safety and health programs
- To promote training in the areas of recognition, avoidance, and the prevention of occupational hazards
- To establish another line of communication whereby the workers can voice their concerns regarding potential hazards and then receive feedback on the status or action taken

---

Summary

What can joint L/M committees accomplish?

- Increased commitment to achieving the organization's goals or mission.
- Improved productivity, safety and health, customer service, and product quality.
- Joint resolution of problems and issues facing the organization.
- Shared responsibility toward and accountability for results and outcomes.
- Better and more constructive relationship between labor and management.
- Enhanced employee morale and job satisfaction.
- Heightened communication and information sharing, which brings all employees into the decision-making process. This helps them understand the mission, goals, and objectives of the organization and fosters employee support of the organization's undertakings.
- Increased job security and compensation.

To make joint L/M committees work, certain actions must occur, and certain procedures must be followed:

- Ensure that upper management supports the joint effort and that this is conveyed to both the union and other company representatives.

- Acknowledge that reservations exist on both sides and try to gradually build trust.
- Keep the committee focused on its goals and mission.
- Strive for a good balance of employee and management representatives who are willing to invest in the process.
- Keep the committee structured; do not allow it to turn into a bull session.
- Remember that the committee is designed to serve all workplace constituencies, not just workers and management.
- Assure that committee leadership is elected or selected by consensus to fill various roles.
- Make decisions fairly and use the consensus process.
- Know and work within the guidelines of federal and state regulations.
- Do not raise issues that really must be addressed at the collective bargaining table; it will only undermine the viability and success of this process.

Further Readings

Reese, C.D. *Joint Labor/Management Committees: A Guide for Committee Members*, Storrs, CT: UConn Press, 1996.

Reese, C.D. *Accident/Incident Prevention Techniques (Second Edition)*. Boca Raton, FL: CRC Press, 2012.

Reese, C.D. *Occupational Health and Safety Management: A Practical Approach (Second Edition)*. Boca Raton, FL: CRC Press/Taylor & Francis Group, 2016.

Reese, C.D. and J.V. Eidson. *Handbook of OSHA Construction Safety & Health (Second Edition)*. Boca Raton, FL: CRC Press, 2006.

16

Workplace Inspections

Workplace audits are inspections that are conducted to evaluate certain aspects of the work environment regarding occupational safety and health (OSH). The use of safety and health audits has been shown to have a positive effect on a company's loss control initiative. In fact, companies who perform safety and health audits have fewer accidents/incidents than companies who do not perform audits. The safety and health audits (inspections), which are often conducted in workplaces, serve a number of evaluative purposes. The following are reasons why audits or inspections are performed:

1. Check compliance with company rules and regulations
2. Check compliance with Occupational Safety and Health Administration (OSHA) rules
3. Determine the safety and health condition of the workplace
4. Determine the safe condition of equipment and machinery
5. Evaluate supervisor's safety and health performance
6. Evaluate workers' safety and health performance
7. Evaluate progress regarding safety and health issues and problems
8. Determine the effectiveness of new processes or procedural changes

Workplace inspections are important for the following reasons:

- They save on cost after something transpires.
- They dwell on the present or potential problems.
- They allow for real-time observations.
- They identify presently existing hazards.
- They can detect existing performance, procedural, and behavioral issues.

Need for an Inspection

First is determining what needs to be audited. What is audited might be a specific occupation (e.g., machinist), task (e.g., welding), topic (e.g., electrical), team (e.g., rescue), operator (e.g., crane operator), part of the work site (e.g., loading/unloading), compliance with an OSHA regulation (e.g., Hazard Communication Standard), or the complete work site. An inspection might be performed if any of the previous lists of activities have unique identifiable hazards, new tasks involved, increased risk potential, changes in job procedures, areas with unique operations, or areas where comparison can be made regarding safety and health factors.

In the process of performing inspections, one might discover hazards in a new process, hazards once a process has been instituted, a need to modify or change processes or procedures, or situational hazards that may not exist at all times. These audits may verify that job procedures are being followed and identify work practices that are both positive and negative. They may also detect exposure factors, both chemical and physical, and determine monitoring and maintenance methods and needs.

At times, inspections are driven by the frequency of injury; potential for injury; the severity of injuries; new or altered equipment, processes, and operations; and excessive waste or damaged equipment. These audits may be continuous, ongoing, planned, periodic, intermittent, or dependent upon specific needs. Audits may also determine employee comprehension of procedures and rules, the effectiveness of workers' training, assess the work climate or perceptions held by workers and others, and evaluate the effectiveness of a supervisor regarding his/her commitment to safety and health.

At many active workplaces, daily site inspections are performed by the supervisor in order to detect hazardous conditions, equipment, materials, or unsafe work practices. At other times, periodic site inspections are conducted by the site safety and health officer. The frequency of inspections is established in the workplace safety and health program. The supervisor, in conjunction with the safety and health officer, determines the required frequency of these inspections, based on the level and complexity of the anticipated activities and on the hazards associated with these activities. In the review of work site conditions/activities, site hazards, and protecting site workers, the inspections should include an evaluation of the effectiveness of the company's safety and health program. The safety and health officer should revise the company's safety and health program as necessary, to ensure the program's continued effectiveness.

Prior to the start of each shift or new activity, a workplace and equipment inspection should take place. This should be done by the workers, crews, supervisor, and other qualified employees. At a minimum, they should check the equipment and materials that they will be using during the operation or shift for damage or defects that could present a safety hazard. In addition, they should check the work area for new or changing site conditions or activities that could also present a safety hazard.

All employees should immediately report any identified hazards to their supervisors. All identified hazardous conditions should be eliminated or controlled immediately. When this is not possible,

1. Interim control measures should be implemented immediately to protect workers.
2. Warning signs should be posted at the location of the hazard.
3. All affected employees should be informed of the location of the hazard and the required interim controls.
4. Permanent control measures should be implemented as soon as possible.

When a supervisor is not sure how to correct an identified hazard, or is not sure if a specific condition presents a hazard, he/she should seek technical assistance from a competent person, a site safety and health officer, or other supervisors or managers. Safety and health audits should be an integral part of the safety and health effort. Anyone conducting a safety and health audit must know the workplace, the procedures or processes being audited, the previous accident history, and the company's policies and operations. This person should also be trained in hazard recognition and interventions regarding safety and health.

When to Inspect?

The supervisor or project inspector must perform daily inspections of active work sites to detect hazardous conditions resulting from equipment or materials or unsafe work practices. The supervisor, inspector, or site safety and health officer must perform periodic inspections of the workplace at a frequency established in the work site's specific safety and health program. The supervisor, in conjunction with the site safety and health officer, should determine the required frequency of these inspections based on the level and complexity of anticipated work activities and on the hazards associated with these activities. In addition to a review of work site conditions and activities, these inspections must include an evaluation of the effectiveness of the work site safety and health program in addressing site hazards and in protecting site workers. The safety and health officer may need to revise the safety and health program as necessary to ensure the program's continued effectiveness. Work crew supervisors, foremen, and employees need to inspect their workplace prior to the start of each work shift or new activity. At a minimum, supervisors and employees should do the following:

1. Check the equipment and materials that they will use during the operation or work shift for damage or defects that could present a safety hazard.
2. Check the work area for new or changing site conditions or activities that could present a safety hazard.
3. Employees shall immediately report identified hazards to their supervisors.

What to Inspect?

The complexity of the work site and the myriad of areas, equipment, tasks, materials, and requirements can make the content of most inspections overwhelming. As you can see, the audit topics below, which could be targeted on a work site, are expansive.

Inspection Topics		
Acids	Forklifts	Safety devices
Aisles	Fumes	Scaffolds
Alarms	Gas cylinders	Shafts
Atmosphere	Gas engines	Shapers
Automobiles	Gases	Shelves
Barrels	Generators	Signs
Barriers	Hand tools	Solvents

(Continued)

Inspection Topics

Boilers	Hard hats	Stairways
Buildings	Hazardous chemical processes	Steam systems
Cabinets	Heavy equipment	Storage facilities
Catwalks	Hoists	Tanks
Caustics	Horns and signals	Transportation equipment
Chemicals	Hoses	Trucks
Compressed gas cylinders	Housekeeping	Unsafe conditions
Confined spaces	Jacks	Unsafe acts
Containers	Ladders	Ventilation
Controls	Lifting	Walkways
Conveyors	Lighting	Walls and floor openings
Cranes	Loads	Warning devices
Docks	Lockout/tagout	Welding and cutting
Doors	Machines	Work permit
Dusts	Materials	Working surfaces
Electrical equipment	Mists	X-rays
Elevators	Noise	
Emergency procedures	Personal protective equipment	
Environmental factors	Personal services and first aid	
Explosives	Piping	
Extinguishers	Platforms	
Fall protection	Power sources	
Fibers	Power tools	
Fire extinguishers	Radiation	
Fire protection	Railroad cars	
Flammables	Respirators	

Safety and health audits/inspections can be done on or for the entire plant (e.g., manufacturing), a department (e.g., quality control), a specific worker unit (e.g., boiler repair), a job or task (e.g., diving), a certain work environment (e.g., confined spaces), a specific piece of equipment (e.g., forklift), a worker performing a task (e.g., power press operator), or prevention of an event (e.g., fire). It is important to select inspections and audits tailored to meet the needs of your company regarding auditing for safety and health concerns.

Types of Inspection Instruments

Instruments can be developed to meet specific needs and specific purposes, and for a variety of topics. The previous is a list of some inspection topics for which inspection instruments can be tailored to meet the company's needs. These instruments can be as simple as a yes/no checklist, an evaluation instrument, or a narrative format, which requires a longer written response.

Summary

In order to prevent accidents on your jobsite, you definitely should develop audits that meet your company's individual needs and include components for your company's particular type of work. Packaged audits that have been developed by others will not be, in all likelihood, very useful for you since each company is unique and has its own hazards and procedures.

Further Readings

Reese, C.D. *Accident/Incident Prevention Techniques (Second Edition)*. Boca Raton, FL: CRC Press, 2012.

Reese, C.D. *Occupational Health and Safety Management (Third Edition)*. Boca Raton, FL: CRC Press, 2016.

Reese, C.D. and J.V. Eidson. *Handbook of OSHA Construction Safety & Health (Second Edition)*. Boca Raton, FL: CRC/Lewis Publishers, 2006.

United States Department of Labor. Occupational Safety and Health Administration, Office of Training and Education. *OSHA Voluntary Compliance Outreach Program: Instructors Reference Manual*. Des Plaines, IL: US Department of Labor, 1993.

Section D

Personnel Involved in Occupational Safety and Health

The final outcomes and successes of occupational safety and health (OSH) are directly related to those individuals involved and how they fill their roles as members of the teams that oversee, conduct, implement, carry out, and fulfill the intent of OSH within a given workplace.

The contents of this section are as follows:

Chapter 17—Management's Commitment and Involvement

Chapter 18—Line Supervisors

Chapter 19—Workers

Chapter 20—Safety Director or Manager

Chapter 21—Safety and Health Professional

Chapter 22—Industrial Hygienist

Chapter 23—Safety and Health Consultant

17

Management's Commitment and Involvement

In order for occupational safety and health (OSH) to function effectively, there must be a complete dedication and support for it. This is why the term *commitment* is really directed at management since it is solely management's responsibility to provide a safe and healthy workplace for its employees. When occupational injuries and illnesses occur, they are considered to be failings within the management system. Management sets the tone for safety and health within the workplace. It does so by demonstrating a commitment. The first item to look for is a safety and health policy that expresses the position of the company in relation to safety and health. This policy must be signed by the president, chief executive officer (CEO), or similar official, not the safety director. This is why the safety and health program must start at the top and work its way down to every level of the company.

A safety and health policy statement clarifies the policy, standardizes safety within the company, provides support for safety, and supports the enforcement of safety and health within the company. It should set forth the purpose and philosophy of the company, delineate the program's goal, assign responsibilities for all company personnel, and be positive in nature. It should be as brief as humanly possible.

It is expected that the top management will sign off on the company's safety and health program. This does not mean that they will develop it, but that they support it. This is another example of commitment.

This is why top management determines to whom safety and health professionals report. It has always been the contention that the higher, the better, since safety and health often needs some teeth when dealing with line management. The president of the company would be ideal. Of course, everyone realizes that due to size or other constraints, this is not always possible.

At one company, the author while sitting with the safety director trying to solve an existing accident problem for the company, the telephone was answered by saying, "Personnel." This told the whole story at once on what the safety director thought his/her job was and also provided an indication as to why they were having problems. The company did not have a true commitment to OSH. Commitment to OSH was only a necessary evil to this company. They had no direction in their program. This is a reason why commitment and actual support of safety and health are necessary.

Management helps to set the goals and objectives of the program and then signs off on them. Nothing can be accomplished in a positive fashion without attainable goals. Goals are the target. Most of us have been asked by others, "What are your goals?" If this cannot be answered, others will view the OSH effort as lost or at best nonmotivated. If goals have not been set, how will the workforce know what is expected of them in relation to OSH? This is why a company should not go through a year without setting some production goals, and logically, from this, OSH goals should be put in place.

Goals and objectives are very important and should be directly observable and measurable. They should be reasonable and attainable. The following are some examples of goals and objectives and the issues faced by those using them:

- Ensure zero fatalities or serious injuries. (This is usually a pie in the sky or unreachable goal for most employers. For example, if you had 25 accidents last year, 0 is probably not possible.)
- Reduce injuries, lost workday accidents, and workers' compensation claims by _____%.
- Prevent damage or destruction to company property or equipment. List real damage and destruction.
- Increase productivity through reduction of injuries by _____%.
- Reduce workers' compensation costs by decreasing the number of claims to _____ or cost by _____%.
- Enhance company's image by working safely. Can you measure this in some way?
- Keep safety a paramount part of workers' daily activities. What are indicators of this? (They are the number of near misses, reports of hazards, or the number of observable unsafe acts.)
- Recognize and reward safe work practices. How is this part of a goal? What could be the measurable outcome of this objective?

Management should develop and implement a set of safety and health rules and policies. Since these rules are developed for the good and welfare of everyone, this is why everyone should obey and follow them at all times, even when it is not convenient. A manager should never enter a work area that requires protective eyewear without the required eyewear. No special favors should exist. After all, the manager is a direct reflection of management's commitment to the company's OSH. Managers are role models who must emulate the company's philosophy and commitment to the safety and health program, to which the top manager has given his support.

It cannot be expected that an OSH program can run without financial assistance. Thus, it is management's responsibility to support safety and health with an adequate budget. Safety and health should have its own allocated budgetary resources. This is the reason that a separate budget for OSH should be developed. In fact, safety and health should be managed like any other component of the company, whether it is research, development, or production. Management commitment should include tough love. This may sound silly, but what is meant is that if management is committed to enforcing its safety and health, to avoid any deviation from the company's rules and regulations and attaining its goals, there must be some form of disciplinary policy and procedure that results in negative consequences for failure to follow or abide by the company's safety and health expectations. See example under Chapter 19, "Workers." This is not just for the workforce but should be evenly applied to all of management and the line supervisors, for any events or times of noncompliance.

Why Roles and Responsibilities?

Everyone is responsible for safety and health and accident/incident prevention. This statement also means that no one in particular is accountable or responsible. Four entities must be accountable and responsible for the OSH effort:

1. All levels of management must demonstrate their commitment to the company's policies and procedures regarding safety and health by their presence, visibility, actions, adherence, and behavior.

2. The person who, by background or experience, has been assigned responsibility and therefore assigned accountability to ensure that the company's safety and health program is adhered to.

3. The supervisor, who models the company's safety personality and is the liaison between management and the worker relevant to the implementation of safety, must be held both responsible and accountable for safety in his/her work area.

4. Employees are responsible for abiding by the company's rules and policies and are accountable for their own behavior, safe or unsafe.

Each of these entities must understand both their responsibilities and accountabilities regarding the safety and health policies and procedures of the company. The sum total of their actual actions and behavior makes up the outward manifestation of the company's culture.

Why Management?

The accountability for workplace safety and health falls on the management. Thus, those who are held accountable are the ones who are responsible for all aspects of the safety and health program. Occupational Safety and Health Administration (OSHA) does not cite workers or the safety and health professional. It cites the company, and the representatives of the company are the management team.

It is interesting to note that nearly 50 employees are injured every minute of a 40-hour workweek and almost 17 employees die each day in the United States. OSHA states that almost one-third of all serious injuries and illnesses stem from overexertion or repetitive trauma, which are expensive injuries. It is important that management views safety and health as an ethical and moral obligation as well as a monetary obligation. Research studies indicate that companies can expect a $4 to $6 return for every dollar invested in safety and health. This is why development of programs and commitment to managing safety and health demonstrates that this not only is a long-term commitment to do the right thing to protect workers from injury and illness. It also proves to be good for the company's bottom line.

Some factors the managers should undertake to demonstrate their commitment to the company's occupational safety and health effort are as follows:

- Communicating with everyone including other peers, supervisor, and employees
- Making decisions and problem solving regarding safety and health
- Disseminating information on hazards, rules, and compliance to maintain a safe workplace
- Inspecting and maintaining a safe workplace for all employees
- Abiding by and enforcing safety and health rules and regulations
- Staying updated on emerging safety and health issues
- Demonstrating openly by deeds and actions support for safety and health
- Using the appropriate resources to provide a safe and healthy workplace

In summation, these are ways management's commitment is effectively conveyed:

- Inspecting facilities, machinery, and safety equipment to identify and correct potential hazards, and to ensure safety regulation compliance.
- Interpreting safety regulations for others interested in industrial safety such as safety engineers, labor representatives, and safety inspectors.
- Maintaining and applying knowledge of current policies, regulations, and industrial processes.
- Maintaining liaisons with outside organizations such as fire departments, mutual aid societies, and rescue teams, so that emergency responses can be facilitated.
- Report or review findings from accident investigations, facility inspections, or environmental testing.

Further Readings

Reese, C.D. *Accident/Incident Prevention Techniques (Second Edition)*. Boca Raton, FL: CRC Press, 2012.
Reese, C.D. *Occupational Health and Safety Management (Third Edition)*. Boca Raton, FL: CRC Press, 2016.
Reese, C.D. and J.V. Eidson. *Handbook of OSHA Construction Safety & Health (Second Edition)*. Boca Raton, FL: CRC/Lewis Publishers, 2006.

18

Line Supervisors

The individual with the most impact on workplace safety and health has to be the line supervisor. Everything that is communicated to the workers comes from the supervisor. The first-line supervisor sets the tone for his/her workplace. The supervisor is the role model for the company, since he/she conveys, implements, supports, and enforces all the company's policies and procedures from production to safety.

Just think of all that the first-line supervisor does:

1. Hires new employees
2. Trains new employees
3. Holds safety meetings
4. Coaches employees on the job
5. Controls quality and quantity
6. Stops a job in progress
7. Takes unsafe tools out of production
8. Investigates accidents
9. Inspects the work area
10. Corrects unsafe conditions and unsafe acts
11. Recommends promotions or demotions
12. Transfers employees in and out of the work area
13. Grants pay raises
14. Issues warnings and administers discipline
15. Reports on probationary employees
16. Suspends and/or discharges
17. Prepares work schedules
18. Delegates work to others
19. Prepares vacation schedules
20. Grants leaves of absence
21. Lays off others for lack of work
22. Processes grievances
23. Authorizes maintenance and repairs
24. Makes suggestions for improvement
25. Discusses problems with management
26. Reduces waste
27. Prepares budget

28. Approves expenditures
29. Fosters employees' morale
30. Motivates workers
31. Reduces turnover

Is it any wonder the success or failure of the company's safety and health program is dependent upon the first-line supervisor? Certainly, everyone would acknowledge that the first-line supervisor is responsible for safety and health within his/her work area. But seldom is the line supervisor evaluated on his/her safety performance in the same manner as his/her production performance. Until the supervisor is held accountable for safety and health in the same manner as for production with equal consequences for poor safety and health performance as for poor production performance, then safety and health will never be a priority with him/her. The value that the supervisor places upon safety and health will always be far less than the value placed upon production. This is why a separate evaluation form for the supervisors' safety and health performance, which can be used to compare safety and health performance of supervisors, will go a long way toward placing equal value on occupational safety and health (OSH).

Further Readings

Reese, C.D. *Office Building Safety and Health*. Boca Raton, FL: CRC Press, 2004.
Reese, C.D. *Accident/Incident Prevention Techniques (Second Edition)*. Boca Raton, FL: CRC Press/Taylor & Francis Group, 2012.
Reese, C.D. *Occupational Health and Safety: A Practical Approach (Third Edition)*. Boca Raton, FL: CRC Press/Taylor & Francis Group, 2016.
Reese, C.D. and J.V. Eidson. *Handbook of OSHA Construction Safety & Health (Second Edition)*. Boca Raton, FL: CRC/Taylor & Francis, 2006.

19

Workers

Everyone is responsible for occupational safety and health (OSH). This statement also means that no one in particular is accountable or responsible. Three entities must be accountable and responsible for accident prevention. These three are as follows:

1. The person who, by background or experience, has been assigned responsibility and therefore assigned accountability to assure that the company's safety and health program is adhered to.
2. The supervisor who models the company's safety personality. The supervisor is the liaison between management and the worker relevant to the implementation of safety. The supervisor must be held both responsible and accountable for safety in his/her work area.
3. The employees who are responsible for abiding by the company's rules and policies and are accountable for their own behavior, safe or unsafe.

Each of the aforementioned entities must understand both their responsibilities and accountabilities regarding the safety and health policies and procedures of the company.

The first two were discussed earlier in this section. It might be wondered why employees (workers) would even be mentioned. The main reason is the nature of their work. OSH is directed toward the work that they are to perform. If they perform their work in a safe and healthy manner, then they will end their workday without death, injury, or illness and return home in the same condition as when they left. This is why the employer has certain expectations of the workforce. If the company fulfills its obligation, then why should employees not shoulder personal responsibility and perform according to expected safety and health behaviors?

Why Workers?

Although workers do not have the control that management has over the workplace, they are still responsible for complying with the company's safety and health policies and procedures. This is why commonly accepted worker responsibilities are expected by employers, and workers are held accountable for their adherence to them.

1. Comply with the Occupational Safety and Health Administration's (OSHA's) regulations and standards.
2. Do not remove, displace, or interfere with the use of any safeguards.
3. Comply with the employer's safety and health rules and policies.
4. Report any hazardous conditions to their supervisor or employer.

5. Report any job-related injuries and illnesses to their supervisor or employer.

6. Report near-miss incidents to their supervisor or employer.

7. Cooperate with the OSHA inspector during inspections when requested to do so.

8. Report to work on time.

9. Wear suitable work clothes.

10. Observe good personal hygiene.

11. Sleeping, gambling, horseplay, fighting, theft, fireworks, and firearms are strictly prohibited on the job, and are grounds for immediate dismissal.

12. Using or being under the influence of alcohol, narcotics, or other drugs or intoxicants is strictly prohibited.

13. Wear your personal protective equipment as prescribed for each task.

14. Housekeeping is everyone's responsibility.

15. Observe "danger," "warning," "caution," and "no smoking" signs and notices.

16. Use and handle equipment, materials, and safety devices with care.

17. Do not leave discharged fire extinguishers in the work areas.

18. Do not expose yourself to dangerous conditions or actions.

19. Do not operate any equipment when you have not been trained.

20. Participate in all safety and health training provided.

21. Attend all safety and toolbox meetings (mandatory).

If a worker elects not to follow rules or to perform work in an unsafe manner, then there should be consequences, which must be strictly enforced by the company and its representatives. Failure to do so negates the authority of the safety and health policies and procedures. Each company should have a discipline policy, which is progressive and stringently enforced. A policy at its simplest is presented as follows:

1. Verbal warning—first offense

2. Written warning—second offense

3. Suspension from work—third offense

4. Dismissal/termination—fourth offense*

This why employees must be held accountable for poor safety performance. As can be seen, each of these three entities has responsibility for safety and health in some form or fashion and should be held strictly accountable for their safety and health performance.

---

* Company reserves the right to immediately dismiss a worker when he/she flagrantly endangers other workers, causes an incident, or causes imminent danger to himself/herself or others.

Further Readings

Reese, C.D. *Accident/Incident Prevention Techniques (Second Edition)*. Boca Raton, FL: CRC Press, 2012.

Reese, C.D. *Occupational Health and Safety Management (Third Edition)*. Boca Raton, FL: CRC Press, 2016.

Reese, C.D. and J.V. Eidson. *Handbook of OSHA Construction Safety & Health (Second Edition)*. Boca Raton, FL: CRC/Lewis Publishers, 2006.

United States Department of Labor, Mine Safety and Health Administration, *Accident Prevention, Safety Manual No. 4*, Revised 1990.

United States Department of Labor. Occupational Safety and Health Administration. Office of Training and Education. *OSHA Voluntary Compliance Outreach Program: Instructors Reference Manual*. Des Plaines, IL: US Department of Labor, 1993.

20

Safety Director or Manager

Is important when managing a company's occupational safety and health (OSH) initiative that someone should be designated as the overseer and responsible party of safety and health. The individual is usually designated as safety manager or director. This individual must both be experienced and have an understanding of the specific hazards that exist in the company's workplace(s). This individual may also be called the safety coordinator or person. No matter the title, his/her responsibilities are varied and wide ranging. Some possible performance expectations may be as follows:

1. Establishing programs for detecting, correcting, or controlling hazardous conditions, toxic environments, and health hazards.

2. Ensuring that proper safeguards and personal protective equipment are available, properly maintained, and properly used.

3. Establishing safety procedures for employees, plant design, plant layout, vendors, outside contractors, and visitors.

4. Establishing safety procedures for purchasing and installation of new equipment and for the purchase and safe storage of hazardous materials.

5. Maintaining an accident recording system to measure the organization's safety performance.

6. Staying abreast of, and advising management on, the current federal, state, and local laws, codes, and standards related to safety and health in the workplace.

7. Carrying out the company's safety obligations as required by law and/or union contract.

8. Conducting investigations of accidents, near misses, and property damage and preparing reports with recommended corrective action.

9. Conducting safety training for all levels of management and newly hired and current employees.

10. Assisting in the formation of both a management and a union/management safety committee (department heads and superintendents) and attending monthly departmental safety committee meetings.

11. Keeping informed on the latest developments in the field of safety such as personal protective equipment, new safety standards, workers' compensation legislation, and new literature pertaining to safety, as well as attending safety seminars and conventions.

12. Maintaining liaison with national, state, and local safety organizations and taking an active role in the activities of such groups.

13. Accompanying Occupational Safety and Health Administration (OSHA) compliance officers during plant inspections and insurance safety professional on audits

and plant surveys. The safety engineer further reviews reports related to these activities and, with management, initiates action for necessary corrections.

14. Distributing the organization's statement of policy as outlined in its organizational manual.

If some facets of the safety effort are not going well, this individual will usually be held accountable even though he/she may not have the authority to rectify the existing problem. Usually, the safety person has a staff position that seldom allows him/her to interfere in any way with the line function of production. Without some authority to impact line functions when necessary, the safety director or manager has little clout as to worksite implementation of the company's safety and health effort. Accountability and responsibility must go beyond this individual. Dan Petersen, noted safety expert, has espoused that what is desired is "safe production," which intimates that the production personnel should be held accountable and responsible for the safe performance of their duties, which includes safety and health.

Further Readings

Petersen, D. *Techniques of Safety Management: A Systems Approach (Third Edition)*. Goshen, NY: Aloray Inc., 1989.

Reese, C.D. *Office Building Safety and Health*. Boca Raton, FL: CRC Press, 2004.

Reese, C.D. *Handbook of Safety and Health for the Service Industry: Volume 1, Industrial Safety and Health for People-Oriented Services*. Boca Raton, FL: CRC Press/Taylor & Francis Group, 2009.

Reese, C.D. *Accident/Incident Prevention Techniques (Second Edition)*. Boca Raton, FL: CRC Press/ Taylor & Francis Group, 2012.

Reese, C.D. *Occupational Health and Safety: A Practical Approach (Third Edition)*. Boca Raton, FL: CRC Press/Taylor & Francis Group, 2016.

Reese, C.D. and J.V. Eidson. *Handbook of OSHA Construction Safety & Health (Second Edition)*. Boca Raton, FL: CRC/Taylor & Francis, 2006.

21

Safety and Health Professional

The safety and health professional may range from the safety director of a company with responsibility for safety and health at a company, facility, or jobsite to an individual with specific expertise. This is why in general terms, an occupational safety and health (OSH) professional is there to protect workers from harm and prevent damage to equipment, property, the environment, and the public. This is accomplished by analysis, design, and implementation of programs to prevent occupationally related injuries and illnesses. Many safety and health professional specialize in specific areas, with expertise in engineering, industrial hygiene, system safety, loss control, and ergonomics, while others handle all facets of OSH. These professionals may work for the public sector, private sector, or government, or as consultants. They are found in a host of industry sectors such as mining, military, service industries, construction, hazardous waste, chemical handling and processing, manufacturing, insurance, transportation, longshoring, agriculture, energy source production, research and development, and a multitude of other domains.

Why are safety and health professionals needed? The first and foremost are the following:

1. To have someone that is responsible for safety and health
2. To have expertise in safety and health
3. To develop needed safety and health and prevention programs
4. To implement all the components of the safety and health initiative
5. To train others and assure all workers are trained in compliance with Occupational Safety and Health Administration (OSHA) and the company's requirements
6. To conduct and supervise safety and health inspections and audits
7. To have a voice or advocate for safety and health

Safety and health requires leadership, and a safety and health professional must be a qualified, knowledgeable individual who fills that role in a company or business. Such a person comes with many titles based upon his/her responsibilities and areas of expertise, such as the following:

- Corporate safety and health manager
- Environmental engineer
- Environmental, health, and safety manager
- Ergonomist
- Hazard control specialist
- Hazardous materials manager
- Health physicist
- Fire protection engineer

- Industrial engineer
- Industrial hygienist
- Industrial psychologist
- Injury prevention specialist
- Loss-control specialist
- Occupational health specialist
- Occupational nurse
- Occupational physician
- Plant safety and health manager
- Risk manager
- Safety director
- Safety engineer
- Safety and health compliance officer
- Safety and health manager
- Safety and health specialist
- Safety inspector
- Safety training specialist
- Security professional

In 1996, the American Society of Safety Engineers (ASSE) published a pamphlet entitled *Scope and Functions of the Professional Safety Position.* This pamphlet provides a superb presentation of why a safety and health professional is important to the arena of OSH by explaining why the safety and health professional must anticipate, recognize, evaluate, control, and communicate safety and health as their main function.

Why is it important to anticipate? The safety and health professional needs to develop methods for anticipating and predicting hazards based upon his/her experience, historical data, and other pertinent sources of information. This will allow for identifying and recognizing hazards in existing and future systems, equipment, products, software, facilities, processes, operations, and procedures during the life expectancy of these various facets of the workplace.

As part of this, he/she must evaluate and assess the probability and severity of loss events and accidents/incidents that may result from the actual or potential hazards.

This requires applying these methods and conducting hazard analyses and interpreting results.

The analysis and interpretation is accomplished by reviewing, with the assistance of specialists where needed, entire systems, processes, and operations, and any subsystems or components, for failure modes, causes, and effects, due to the following:

- System, subsystem, or component failures.
- Human error.
- Incomplete or faulty decision making, judgments, or administrative actions.
- Weaknesses in proposed or existing policies, directives, objectives, or practices.

It is important to review, compile, analyze, and interpret data from accident and loss-event reports and other sources regarding injuries, illnesses, property damage, environmental effects, or public impacts. This is why a safety and health professional must identify causes, trends, and relationships to ensure completeness, accuracy and validity of required information; evaluate the effectiveness of classification schemes and data collection methods; and initiate investigations.

The duty of the safety and health professional is to provide advice and counsel about compliance with safety, health, and environmental laws, codes, regulations, and standards, including conducting research studies of existing or potential safety and health problems and issues.

It is also expected that he/she will determine the need for surveys and appraisals that help identify conditions or practices affecting safety and health, including those that require the services of specialists, such as physicians, health physicists, industrial hygienists, fire protection engineers, design and process engineers, ergonomists, risk managers, environmental professionals, psychologists, and others while assessing environments, tasks, and other elements to ensure that physiological and psychological capabilities, capacities, and limits of humans are not exceeded.

Development of hazard control designs, methods, procedures, and programs is another reason why a safety and health professional is needed. They must formulate and prescribe engineering or administrative controls, preferably before exposures, accidents, and loss events occur; eliminate hazards and causes of exposures, accidents, and loss events; and reduce the probability or severity of injuries, illnesses, losses, or environmental damage from potential exposures, accidents, and loss events when hazards cannot be eliminated.

Such a mandate entails developing methods that integrate safety performance into the goals, operations, and productivity of organizations and their management and into systems, processes, and operations or their components. This also includes the development of safety, health, and environmental policies, procedures, codes, and standards for integration into operational policies of organizations, unit operations, purchasing, and contracting.

The safety and health professional needs to consult with and advise individuals and participate on teams while engaged in planning, design, development, and installation or implementation of systems or programs involving hazard controls. This should encompass engaging in planning, design, development, fabrication, testing, packaging, and distribution of products or services related to safety requirements and application of safety principles that will maximize product safety.

The safety and health professional should use behavior-based safety techniques by advising and assisting human resources specialists when applying hazard analysis results or dealing with the capabilities and limitations of personnel.

If the previous paragraphs do not justify why there is need for a safety and health professional, the following duties indicate why not everyone or anyone can be expected to fill this position. They are expected to stay current with technological developments, laws, regulations, standards, codes, products, methods, and practices related to hazard controls. They implement, administer, and advise others on hazard controls and hazard control programs, preparing reports that communicate valid and comprehensive recommendations for hazard controls that are based on analysis and interpretation of accident, exposure, loss-event, and other data. The must also use written and graphic materials, presentations, and other communication media to recommend hazard controls and hazard control policies, procedures, and programs to decision-making personnel.

Other responsibilities include directing or assisting in planning and developing educational and training materials or courses; conducting or assisting with courses related to designs, policies, procedures, and programs involving hazard recognition and control; advising others about hazards, hazard controls, relative risk, and related safety matters when they are communicating with the media, community, and public; and finally, managing and implementing hazard controls and hazard control programs that are within the duties of their profession in the safety and health position.

In order to evaluate the effectiveness of the safety and health professional, one will need to measure, audit, and evaluate the effectiveness of hazard controls and hazard control programs. The safety and health professional will always be pressed to justify the impact and effectiveness of OSH. He/she will need to establish and implement techniques that involve risk analysis, cost, cost–benefit analysis, work sampling, loss rate, and similar methodologies for periodic and systematic evaluation of hazard control and hazard control program effectiveness. He/she also needs to develop methods to evaluate the costs and effectiveness of hazard controls and programs and measure the contribution of components of systems, organizations, processes, and operations toward the overall effectiveness.

With the above responsibilities, the safety and health professional will be charged with providing results of evaluation assessments, including recommended adjustments and changes to hazard controls or hazard control programs, to individuals or organizations responsible for their management and implementation. This is to be accomplished by directing, developing, or helping to develop management accountability and audit programs that assess safety performance of entire systems, organizations, processes, and operations or their components and involve both deterrents and incentives.

The safety and health professional is the purveyor of safety and health. Thus, he/she must develop and utilize communication techniques that facilitate the company's or business's message regarding safety and health within their workplace. This may be accomplished in a myriad of ways including, but not limited to, direct contact, message display, e-mails, training, safety and health talks, and other creative approaches.

If the safety and health professionals perform their expected functions, they will have the data and justification to develop, implement, and secure the budget needed to support the safety and health effort. Facts, concrete data, and the savings accomplished by the safety and health initiative will always be more supportive of safety and health than assumptions and vague guesses.

Further Readings

American Society of Safety Engineers (1996). *Scope and Functions of the Professional Safety Position*. Des Plaines, IL: American Society of Safety Engineers.

Petersen, Dan, *Techniques of Safety Management: A Systems Approach (Third Edition)*. Goshen, NY: Aloray Inc., 1989.

Reese, C.D. *Occupational Health and Safety Management: A Practical Approach (Second Edition)*. Boca Raton, FL: CRC Press/Taylor & Francis, 2009.

Reese, C.D. *Accident/Incident Prevention Techniques (Second Edition)*. Boca Raton, FL: CRC Press/Taylor & Francis, 2011.

22

Industrial Hygienist

The following is one reason why an employer may have a need for an industrial hygienist (IH). The Occupational Safety and Health Act (OSHAct) has brought a restructuring of programs and activities relating to safeguarding the health of workers. Uniform occupational health regulations now apply to all businesses engaged in commerce, regardless of their locations within the jurisdiction. Nearly every employer is required to implement some element of an industrial hygiene or occupational health or hazard communication program, to be responsive to the Occupational Safety and Health Administration (OSHA) and the OSHAct and its health regulations.

In carrying out such an assessment regarding health-related stressors, a company may not have the expertise or even the equipment that might be needed to do a viable and proper assessment to protect the health of its workforce. From the information, it can be determined when a company has limitations as well as why an IH is needed. Industrial hygiene has been defined as "that science or art devoted to the anticipation, recognition, evaluation, and control of those environmental factors or stresses, arising in or from the workplace, which may cause sickness, impaired health and well-being, or significant discomfort and inefficiency among workers or among the citizens of the community."

The IH, although trained in engineering, physics, chemistry, or biology, has acquired, by undergraduate or postgraduate study and experience, the knowledge of the effects on health of chemical and physical agents under various levels of exposure. The IH is involved with the monitoring and analytical methods required to detect the extent of exposure and the engineering and other methods used for hazard control. All of this preparation is needed and desirable in order to address the myriad of workplace stressors that impact health in the workplace, such as the following:

- Physical stressors—radiation, noise, vibration, temperature extremes, and pressure issues
- Biological stressors—vermin, insects, molds, fungi, viruses, and bacteria
- Ergonomic stressors—design of work area and stations, tool design, material handling, repetitive motion, awkward work positions, lighting, and visual acuity
- Chemical stressors—poisons, toxins, dust, vapors, fumes, and mists

Why an Industrial Hygienist?

An IH is an individual having a college or university degree(s) in engineering, chemistry, physics, medicine, or related physical and biological sciences who has also received specialized training in identification/recognition, evaluation, and control of workplace stressors and has competence in industrial hygiene. The specialized studies and training must be sufficient so that the individual is able to do the following:

- Anticipate and recognize the environmental factors and understand their effects on people and their well-being
- Evaluate, on the basis of experience and with the aid of quantitative measurement techniques, the magnitude of these stressors in terms of the stressor's ability to impair human health and well-being
- Prescribe methods to eliminate, control, or reduce such stresses when necessary to diminish their effects

The IH is a very diversely trained individual. IHs have a strong background in chemistry. They must have a background in engineering, biological sciences, and behavioral/social sciences. The IH has specific training in environmental sampling and is prepared to make recommendations involving solutions and controls for environmental factors that can cause health effects in the workplace. This is why IHs are probably the most highly trained individual on safety and health found and involved in the workplace. This is why physicians were initially threatened by industrial hygiene. But, the IH professional's role in health is recognized by the unique certification.

The IH can perform the following operations, which demonstrate why they are important to safety and health in the workplace:

- Identify potential risk factors that can create health effects in your workforce.
- Evaluate the chemicals that you are using and make recommendations on controls.
- Select and conduct sampling methods for chemical and other environmental factors.
- Recommend the appropriate personal protective equipment. An IH should definitely be the individual selecting and recommending the type of respirator that is needed in the workplace.
- If you need more ventilation in the workplace, IHs are trained to assist and advise you.
- If you are faced with ergonomic issues, the IH has the type of background that can help you solve them.
- If biological agents exist in your workplace, the IH can help identify, evaluate, and develop controls for you.
- IHs can address other diverse health hazards faced in the workplace such as radiation, temperature extremes, vibration, and noise issues, to name a few.

When an Industrial Hygienist Is Needed

Employers need to have the knowledge and courage to realize when they need to call an IH for help. IHs have very special and specific training related to workplace environmental evaluations and assessments as well as the ability to make recommendations on controlling workplace hazards.

An IH concerned about exposure hazards associated with the workplace must be familiar with the various activities and processes that a specific employer has. The classic approach of recognition, evaluation, and control strategies used by IH applies to all industries. Sometimes, exposures can be attributed to the job. For example, for a worker using a solvent to clean a piece of mechanical equipment, the IH may need to investigate organic vapor exposure, correct personal protective equipment use, surrounding environment, and possibly personal hygiene conditions.

Hazards involving normal work activities can usually be predicted by a trained IH. It is, however, very unpredictable how much airborne exposure a worker is subjected to from a particular source. Many times, the same type of work conducted at one site is much different from an exposure condition at another. Inside exposures will remain more constant than outside, where wind and weather conditions play a major role. For example, asbestos abatement work that is conducted in a controlled atmosphere inside should remain fairly constant if work practices such as negative air filtration are used and surfaces are wetted properly. Conversely, work on an asbestos roof on the outside, even though there is a difference in the type of asbestos, will depend more on weather conditions. Work practices such as location of the worker in relationship to the wind (upstream or downstream) and how intact the shingles are as they are removed also play an important part in overall exposure. The more broken up they are, the more likely that an asbestos exposure will result. Although inside exposures sometimes can vary vastly with the size of an area and individual work practices, it is not usually expected to be that way.

If the airborne exposure is to be determined for a particular job, the IH must be prepared to monitor quickly. The next day may be too late. Concentrations usually need to be high to find time weighted average (TWA) that exceed OSHA permissible exposure limit (PEL). More often than not, the construction worker is not conducting the same job for an 8-hour period. Many tasks are usually required to accomplish a day's work, which also makes it difficult to evaluate a particular hazard. A worker welding, cutting, and burning all day on an outside project such as a painted bridge may have no exposure or wind up in the hospital undergoing chelation therapy with a blood lead level in the hundreds. Many variables affect the potential and real exposure levels, such as work habits, weather, and type of paint on the steel, as well as personal protective equipment used.

It is most appropriate to consult an IH when selecting personal protective equipment for a specific use, such as which gloves are best for use with certain chemicals and which respirator should be used for exposure to a specific chemical. This is why the IH is the only one who has the training and experience to determine the risk of exposure, the environmental sampling that is needed, the sampling techniques to use, and the controls that should be in place to prevent further exposure.

Further Readings

Reese, C.D. *Accident/Incident Prevention Techniques (Second Edition)*. Boca Raton, FL: CRC Press/Taylor & Francis, 2012.

Reese, C.D. *Occupational Health and Safety: A Practical Approach (Third Edition)*. Boca Raton, FL: CRC Press/Taylor & Francis Group, 2016.

Reese, C.D. and J.V. Eidson. *Handbook of OSHA Construction Safety & Health (Second Edition)*. Boca Raton, FL: CRC Press/Taylor & Francis, 2006.

United States Department of Labor. Occupational Safety and Health Administration, Office of Training and Education. *OSHA Voluntary Compliance Outreach Program: Instructors Reference Manual*. Des Plaines, IL: Department of Labor, 1993.

United States Department of Labor. Occupational Safety and Health Administration, Office of Training and Education. *Manual for Trainer Course in OSHA Standards for the General Industry*. Des Plaines, IL: Department of Labor, 2001.

23

Safety and Health Consultant

From time to time, each of us is faced with problems or issues that surpass our own training, experience, or expertise. An educated person is one who recognizes that he/she needs help. There are always plenty of individuals ready and willing to provide advice and help. But, make sure that the consultant is competent help that can truly and meaningfully provide the assistance that is desperately needed. This is why when the company realizes that they do not have the knowledge to do what is required, it is good to be confident enough in themselves to understand that limitations exist. This is the appropriate time to find a consultant to be of help to the company. The company will want to make sure that they are getting their money's worth, which is only good fiscal sense.

Why a Need for a Consultant?

The consultant that is needed may be a specialist in a particular area of safety and health such as an ergonomist or an engineer who can help with redesigning issues. No matter the person you need, the company must proceed in an organized fashion in selecting that individual and finally obtaining a solution to the problem. Companies may use occupational safety and health (OSH) consultants for a number of reasons, which include the following:

- Identification of new safety or health problems that require technical, professional resources beyond what is available in the company
- A management initiative to redesign, streamline, and enhance current safety and health processes and programs
- A directive to outsource noncore company functions
- Regulatory-driven regulations requiring new and/or additional compliance measures
- Correcting deficiencies in the safety and health program

A consultant will likely draw on a wide and diverse experience base in helping to address problems and issues. He/she is not influenced by politics and allegiance and is therefore in a better position to make objective decisions. The professional consultant strives to provide cost-effective solutions since his/her repeat business is based upon performance and professional reputation. Also, consultants are temporary employees, and the usual personnel issues are not applicable to them. Consultants work at the times of the day, week, or month when the company has a need and not at their convenience. Most consultants have become qualified to perform these services for you through either education or experience.

In determining the need for a consultant, the company will want to consider whether using a consultant will be cost-effective, faster, or more productive. Companies may also find the necessity for a consultant when they feel a need for outside advice, access to special instrumentation, an unbiased opinion or solution, or an assessment that supports the initial solution.

A consultant can address many of the company's safety and health issues. He/she can also act as an expert witness during legal actions. But primarily, the consultant is hired to solve a problem. Thus, the consultant must be able to identify and define the existing problem and then provide appropriate solutions to your problem. Consultants usually have a wide array of resources and professional contacts, which they can access in order to assist in solving the problem.

To get names and recommendations of potential consultants, contact professional organizations, colleagues, and insurance companies, who often employ loss-control or safety and health personnel. Furthermore, do not overlook local colleges and universities; many times, they also provide consultative services. In selecting a consultant, make sure to ask for the following:

1. A complete resume, which provides you with the consultant's formal education, as well as his/her years of experience
2. A listing of previous clients and permission to contact them
3. The length of time the individual has been a consultant and his/her current status regarding existing business obligations
4. Verification of professional training
5. Documentation of qualifications, such as registered professional engineer, certified safety professional, or certified industrial hygienist
6. A listing of the consultant's memberships in professional associations
7. Any areas of specialization and ownership or access to equipment and certified testing laboratories

Interviewing a Consultant

It is important that proper interviews be conducted of the consultants who have made the final cut. This is why the following items should be part of the evaluation and ranking of the potential consultants:

- Review credentials and qualifications to determine their experience and expertise. Safety and health experts should hold proper credentials.
- Does the consultant have experience with the company's type of problem or project?
- What procedures does the consultant do, and what does he/she subcontract?
- Review the scope of services to determine if the consultant has expertise for the company's requirements.
- Is the consultant's staff trained to properly conduct the scope of work?
- Has the consultant had other projects similar to the company's problem? Ask for references from previous clients that had similar needs. Has he/she had repeat business?

- Does the consultant have to come up to speed, and is there a learning curve?
- Ask for a list of subcontractors and check their performance.
- Does the consultant have professional, compensation, and liability insurance?
- How does the consultant tend to bill for project? Are charges *per diem*, direct personnel cost (time and materials), cost plus fixed payment, fixed lump sum, or percent of cost?
- Is the consultant a member of trade or professional organizations, and what is their status in these organizations?
- Carefully and thoroughly evaluate all references that the consultant provides.
- What is the consultant's present workload?
- Are the consultants and their staff trained in safety and health procedures in accordance with the Occupational Safety and Health Administration (OSHA)?
- Does the consultant have the equipment and testing instruments to carry out any needed procedures?
- Has the consultant met expectations, fulfilled the requirement, met deadlines, and provided a professional and accurate product or deliverables in the past?
- Are there any conflicts of interest?
- Does the consultant have the resources to deliver on the requirements in the scope of work?
- Is the consultant financially stable?
- Do the work practices and testing follow nationally recognized best practice standards?
- Is there a quality control facet, and are testing laboratories certified?
- Will legal representation be needed to safeguard the company's confidential data or trade secrets?

This is an important decision that must be made since a poor decision could have disastrous consequences for the company's safety and health program.

Why a Scope of Work?

The most important document in securing a consultant is the scope of work, which provides some guarantee of attaining the expected outcome when evaluating the consultant. The scope of work should include a description of the needs or project, the company's expectations, any unusual challenges for the consultant, a timeline for completion of the project, the provision for a detailed financial estimate for completing the project, and the expected outputs from the effort such as reports, data, results of testing, or other specified outcomes. The methods and action plan should be part of the response to the scope of work and the resources, manpower, and specialized equipment needed to complete the requirements of the scope of work.

The consultant submitting a proposal for the scope of work might benefit from a visit to the site to consider the problem prior to providing a response to the scope of work. Be careful of proposals that appear overly optimistic, overreaching, extremely low bids, unrealistic timelines, conflicts of interest, hard-sell approaches, minimizing or maximizing of potential technical or legal problems, or unsound approaches to solving the problem.

Although a verbal agreement may seem fine, it is recommended that a written agreement that spells out how much the service is going to cost or the maximum number of hours that the company will support be included in the signed contract. The company should provide a scope of work that sets out the steps that will be followed in order to solve the problem. It is appropriate to include, in this written document, the output expectations from the process. A minimum output would be a written report, but the company may not want to pay for this. If not, a verbal report would be the output in the agreement. All expected outputs should be part of this agreement. This may include drawings, step-by-step procedures, and follow-up. The company needs to get all the information required to solve the problem, based upon the consultant's recommendations. The company may have to construct, implement, or redesign equipment, processes, or procedures. Unless some sort of a protective clause is within the agreement, there is no guarantee that the consultant's recommendations will fix the problem. If a guarantee is desired, expect to pay more since the consultant will be responsible for the implementation of his/her recommendations.

If the process of using a consultant is to work, it must be in full cooperation with the company; thus, the consultant can fail if he/she is not an integral part of the company's team. For this reason, it is important to provide all the information needed. Clearly define the problem. Set objectives. Agree on realistic requirements. Get a clear and complete proposal from the consultant. Review the progress from time to time, and stay within the scope of the problem. The consultant can only be as effective as the company allows him/her to be. The company must do its part to assure success. Make sure that a cost–benefit analysis is performed before pursuing a consultant. Agree on cost and fees prior to starting.

The consultant is not a friend. The company has hired him/her based upon an evaluation of him/her that this is the best person to help solve the problem that the company cannot solve, does not have time to solve, or does not have the resources to solve. The consultant should define the problem, analyze it, and make recommendations for a solution. The consultant can be made a part of your team by providing him/her full information and support. This will assure that the greatest value is reaped from the investment in him/her.

The Hiring Process

This hiring of a consultant is an important business decision. This is why some primary steps for employing a consultant are critical to the process.

1. The company must know what it wants from the beginning.
2. A scope of work should be put in writing.
3. Use a bid process for selecting the consultant.
4. Delineate the time line for completion of work.
5. Decide whom the consultant will be reporting to.
6. Know the company's budget limitations (develop a formal budget).
7. Make sure that a written contract exists.

8. Have the consultant give his/her protocol for completing the work.

9. Discuss the evaluative approach being used by the consultant prior to the contract.

10. Have mechanisms in place to hold the consultant accountable for not completing the agreed-upon work.

11. Establish penalties for not meeting preset schedules.

12. Have a tracking system to evaluate the work's progress.

13. Be consistent in the expectations because each time the scope of work is changed or delays inhibit progress, it will cost extra.

14. Follow good business practices and maintain a professional relationship.

Before selecting a consultant, the official in charge should do his/her homework. Make sure the consultant can service the company's geographic area or multiple site locations depending upon its needs. Hiring a consultant is not an easy process. Assure that the job is not too large for one consultant and may require more than what a friend down the street can offer. Although the company may be familiar with the consultant from a social relationship, this does not mean that he/she can and will provide the service that is needed, unless the company is familiar with his/her work from previous efforts or work done for the company. The company needs to consider what happens in case of illness, disability, or death of the consultant. A contingency plan should exist as to how the work will be continued if some unfortunate event transpires. If special equipment is needed for sampling, for example, visit the consultant's office to assure that they have the necessary resources to accomplish what they are committing to do. This is why hiring a consultant should be undertaken as a business process.

In summary, check the background of the consultant, which should include the consultant's education, industrial experience, membership in professional organizations, and any certification. This is just a starting point. Procure and check the consultant's referrals. The more invested in a consultant, the more referrals and the more the amount of homework needs to be done. Obtain copies of previous contracts and reports and review them for quality, professionalism, and depth of information. A responsible consultant will be happy to provide samples of his/her work.

Remember that company personnel will have to work with the consultant. The chemistry between the consultant and the company's team needs to be there. The attitude of the consultant may be as important as his/her abilities, since he/she must work within your company's framework and with others and yourself. The consultant's inability to communicate effectively can doom the project.

Summary

When a qualified and experienced safety and health consultant is selected, the following are much more likely:

- Money and time will be saved.

- Projects will have to be done only once, because they will be done correctly the first time.

- Relationships with regulatory agencies will be beneficial and cooperative.
- Essential reporting to agencies will be completed in a timely manner.
- The client, the consultant, and the regulatory staff will be satisfied with the final outcome.
- Projects will be completed safely, and unexpected events will be minimized.

Consulting services are an important function in providing assistance relevant to safety and health issues. A consultant is always willing to provide safety and health services. But, depending upon the urgency of the situation or the types of safety and health services needed, the cost may be quite high. Nevertheless, these costs are often justifiable because it is sometimes difficult to secure a consultant who has the expertise needed to assist with specific problems. Remember that the low bid from a consultant may not be in the company's best interest if he/she does not have the expertise to provide the solution needed. For a lower price, the company may get a less qualified consultant. Once a consultant is selected, a contractual agreement needs to be developed. The agreement should delineate the services to be provided, the outcomes or products to be attained, and the follow-up obligations required.

Most consultants are skilled safety and health professionals who bring to the table years of experience and unique expertise. The service that they offer is well worth the price. Using a planned approach will assure that the company is getting the consultant that is needed for the job. But, be cautious, as there are some individuals out there who are less scrupulous and will deceive the company and be happy to take its money, waste time, and leave the company with an inferior product or no product. As Linda F. Johnson advises in *Choosing a Safety Consultant*, "As in all business decisions, do your homework."

Further Readings

Johnson, L.F. "Choosing a Safety Consultant." *Occupational Safety and Health,* July 1999.

Reese, C.D. and J.V. Eidson. *Handbook of OSHA Construction Safety & Health (Second Edition).* Boca Raton, FL: CRC/Lewis Publishers, 1999.

Section E

People-Related Issues

It is interesting to note that most of what is done in life is accomplished with others and through others. It no wonder that our success is dependent on understanding each other's actions and how they apply to everyday life. The principle of working with others is dependent upon how well our communications are sent and received by those in our workplace. Suffice it to say that these principles are as important to conveying and fostering occupational safety and health (OSH) in the work environment as they are to any other part of our life, and using these elements will accomplish a great deal with regards to obtaining a safe and healthy workplace.

The contents of this section are as follows:

Chapter 24—Motivating Safety and Health

Chapter 25—Behavior-Based Safety

Chapter 26—Safety Culture

Chapter 27—Communicating Safety and Health

Chapter 28—Bullying

Chapter 29—Safety (Toolbox) Talks

Chapter 30—Incentives/Rewards

24

Motivating Safety and Health

It is always interesting that while working with employers as well as safety and health professionals, there is such a time and effort investment in developing a safety and health initiative. At times, the cost is sizable. But when motivating employees is mentioned, employers are looking for a quick fix. Seldom do they take time to plan, apply principles of behavior, or invest reasonable amounts of money to motivate employees to work in a safe manner. This seems even odder when one considers that all workplaces are made up of employees and these are the individuals at whom the safety and health effort is aimed. Many times, employers and others say, "What can we give them?" Often, when given a reward or incentive, employees do not even understand why they are receiving it. What most employers and others are asking is, "What is the quick fix?" They think that they can easily address such a complex subject as human behavior by throwing some money at it or giving employees a trinket. It seems reasonable to assert that much more effort should be invested in "getting safety" than usually occurs.

Employers take little, if any, time to think about motivating safety and health. This certainly seems like an oversight when data indicate that 85% to 90% of accident are likely the result of unsafe behavior (acts). With this realization, it seems beyond logic that employers do not pay more attention to motivation in the workplace, especially related to occupational safety and health (OSH) performance.

A word of caution: paying attention to developing motivational approaches to safety and health will be to no avail unless all of the other components discussed in this book are addressed first.

It cannot be expected that workers will be motivated toward safety and health without the foundation of a safety and health program in place. Workers cannot be motivated without knowing what they are expected to be motivated about (or even why they should care about being motivated) if the company hasn't made the effort to define and direct the performance desired regarding safety and health. Much of the development and planning for implementing a motivational approach will have already been completed if the guidelines provided in this book are implemented. These guidelines will have the needed directions, goals, policies, and procedures in place. Another word of caution is that there is no foolproof motivation plan that is fail-safe and assures employers of achieving the results that they desire. As we all know, the most difficult tasks faced in the workplace are those where people are involved.

The reason that most failures occur in trying to get the type of motivation desired is that attempts to change people's values, which are set in early life, or change their attitudes, which are an integral part of their personalities, are met with resistance. Both values and attitudes are not measurable or easily observable and are accepted or rejected based upon an individual's set of values and attitudes. The best that can be hoped for is to change an individual's behavior, which is observable and measurable. Over time, the workers' attitudes may change, or their values may be altered by the employer's motivational attempt, but that is not as important as obtaining safe and healthy work behavior. It is imperative that employers motivate workers to exhibit a behavior that will keep them safe and healthy in the performance of their jobs.

Setting the Stage

The aim is to provide a basis that will allow employers to motivate themselves and their employees. When discussing motivating oneself or others, it is the search for the perfect blueprint. The magic formula for motivation is perceived to be composed of plans, tricks, gimmicks, or inducements. This may be the case in some instances, or for some situations, but this is not the panacea.

The human aspect of safety and health is often brushed aside in industrial settings. After all, each workplace is different, and the fundamental principles of motivating and dealing with construction workers and office workers are, therefore, different also. Granted, people are unique, but all workers have basic needs to be addressed and fulfilled in order for them to work effectively and productively. The principles and examples in this part have been used for miners as well as office workers and are applicable to any group of workers. It is up to employers to use their ingenuity and creativity to develop an effective approach the will motivate the workers that are found in their workplace.

The management of people is directly related to understanding the basic principles of motivation. Obtaining good safety and health behavior and work practices can be directly attributable to how effectively an employer applies good principles of motivation. The intention of this part is to provide a practical and insightful view of the workings of motivation in the work or occupational environment. One must remember that many theories related to motivation exist. Using the principles discussed in this part is not a surefire guarantee that employers will be able to motivate workers, but if something is not done, then certainly employers will not attain the motivation desired in the workforce. Most of the time, managers, contractors, and safety and health professionals want a Band-Aid solution to their motivational problems in the workplace.

Motivation is a somewhat imprecise science and undertaking. No guarantees for success exist. Each workplace needs to pay close attention to the motivational needs of the individuals who work there or there will be no progress in attaining the desired safe behaviors, and the employer will probably see a decrease in morale, since morale is as difficult to define as motivation.

There are many legitimate reasons for trying to motivate workers and employees. These reasons may be as simple as trying to get an employee to work safely or as complex as fostering safe work teams. It is very challenging to instill motivation where motivation does not presently exist. While employers should hesitate to use workers as behavioral guinea pigs, workers are no different from anyone else when it comes to motivational issues. Motivating oneself and others is usually tackled because there is caring about someone or some group and a desire to see them accomplish a goal or conquer more than they ever thought was possible. It is rewarding to see a group of workers attain the goal of working 1,000,000 man-hours without an accident and go on from there.

Although all the principles espoused here can apply to families, relationships, friendships, coworkers, teams, or employees, most of the successes discussed are those relevant to the workplace. This is, of course, because the majority of our adult life is spent in the workplace.

Defining Motivation

Motivation, in the broadest sense, is self-motivation, complex, and either needs driven or value driven. Someone once stated that he believed *hope* was the secret ingredient to a person being motivated (the hope to accomplish a goal or a dream, or attain a need); there is reason to support this theory. But, possibly a better definition is that "motivation presumes valuing, and values are learned behaviors; thus, motivation, at least in part, is learned and can be taught" (Frymier, 1968). This definition provides us with the encouragement we need in order to go forward and achieve motivation for ourselves and others.

If we want to be successful, we must believe that we can teach someone to be motivated toward specific outcomes (goals) or, at the least, be able to alter some unwanted behavior. On the other hand, we should not want to completely manipulate an individual to the point that he/she responds blindly to our motivational efforts. If that occurred, we would lose that most important portion of a person, the unique human will.

Thus, motivation is internal. We cannot directly observe or measure it, but a glimpse of the results may be observed when we see a positive change take place in behavior. Such a change might be something as simple as a worker wearing protective eyewear or something as far reaching as going a full year without an accident or injury. By observing these outwardly manifested behaviors, you can then be encouraged when you see even the smallest of successes that are related to your motivational techniques.

Principles of Motivation

Goals are an integral part of the motivational process and tend to structure the environment in which motivation takes place. The environment in which we find ourselves is many times the springboard to the overall motivational process. An employer may be fortunate enough to accidentally step into a high-energy motivational environment. On the other hand, one may find oneself in an environment that is not at all conducive to motivating others, and thereby, it is very difficult to attain any desired goals. If this is the case, a need to make a change in the physical environment or possibly even make a change in the work atmosphere (i.e., allowing more independence of individuals in the decision-making process) may exist.

The changing of the environment may not affect all individuals in the same way. It has been said that there are three certainties that can be stated regarding people. Those certainties are, "People are unique, people are unique, and people are unique!" Since people are unique, what motivates one person may be demotivating to another.

If as a motivator, one has the responsibility of trying to motivate an individual or group, he/she will need to address their motivational needs. Employees fall along a continuum— some need little motivation from anyone, and others need constant attention. It is unrealistic to expect all of them to achieve a desired level of expectations. The quality of leadership will be the determining factor of success with these workers.

Why are some leaders more motivational than others? What are the unique talents that these dynamic leaders possess? Some people believe that these individuals were born to be leaders. Most of us believe not that they are just born leaders but that they are individuals who possess a set of talents and have chosen to develop those talents to the maximum.

These talents are developed because they have the burning desire (goal) to become leaders and those desires motivate them to learn the necessary skills.

To be a successful motivational leader, there must be some sort of plan that will get the individual or group from point A to point B. This plan should include the desired goals and objectives, levels of expectations, mechanisms for communication, valuative procedures and techniques for reinforcement, feedback, rewards, and incentives. Any motivational plan is a dynamic tool that must be flexible enough to address changes, which may occur over a period of time, and take into consideration the universality of people and situations. These plans can use a variety of techniques and gadgetry to facilitate the final desired outcomes or performances, which lead to a safer and healthier workplace.

The Motivational Environment

Everything that surrounds us is part of the motivational environment. Depending upon our environment, individuals or groups are motivated differently at a given point in time. The motivation population actually exists in what can be called "micromotivational" environments. These microenvironments make up the sum total of the motivational environment and comprise the work environment, family environment, social environment, team environment, peer environment, and even a nonfunctional environment.

Any one or all of these microenvironments can have an impact on the other. The negative impact of one of a person's microenvironments may cause that person to also react negatively in another one of his/her environments; this can happen even when the environment is, in itself, a positive one. For example, if an individual has problems at home, it may, and many times does, cause that highly motivated employee to become less safety conscious or productive at work.

In illustrating the complexity of this issue, let's think for a moment about problem employees. Many times, these employees ask to be moved to a different job because they are either dissatisfied or performing poorly in their current job. Amazingly, once the reassignment is made, their performance vastly improves. It is almost as if this worker becomes a different individual. When they are put into a new and different environment, they get a new spark, and the new environment becomes their positive motivator; they've been revitalized! Many individuals do not like change, but all of us react to and are energized both negatively and positively by change. So changes in the motivational approach need to occur when motivation begins to wane.

Structuring the Motivational Environment

It takes some degree of organization and commitment to structure an environment where workers will be motivated to perform their work in a safe and healthy manner. The safety and health environment must have a foundation. A written safety and health program is the key component in providing and structuring that foundation. This written program should set the tone for safety and health within the work environment.

There are some keys to motivating safety and health. A listing of them is as follows:

1. Explanations
2. Goals
3. Needs
4. Feedback
5. Consequences
6. Reinforcement
7. Involvement
8. Self-monitoring
9. Rewards

First, the motivator must explain and clarify the safety and health performance expectations. It cannot be assumed that workers know what is expected of them unless they are told. These expectations need to be concise and consistent. Tell people how you hope to achieve expectations, and do not fail to ask for advice on how to attain these expectations from everyone in the workforce. Always remember that people have a definite need to know. They certainly like to know what is going on.

Secondly, a way to keep the expectations out in front regarding safety and health is to establish attainable and reachable safety and health goals. Goals that are understandable by all are much better motivators than ones that workers do not understand. It is good to involve those who will be impacted by the goal in the development of that goal. A goal to reduce our injury rate by 20% may sound fine to you, but many workers do not know what an injury rate is or the components that go into calculating it. A better goal would be to keep the number of injuries below 10 per month. This can be easily tracked and counted on a monthly basis. The progress toward this goal can be posted regularly. All of us are goal driven, whether we recognize it or not. Most of what an organization, team, or individual accomplishes is the result of a set goal.

Thirdly, needs are strong motivators of individuals. When a specific need exists, such as food, security, personal needs, or internal achievement, most individuals will move mountains to attain it.

A fourth key to motivation is providing feedback. If the number of injuries each month is posted, workers are being given feedback on the progress toward that goal. Everyone needs to know how they are doing in order to maintain a focus and motivation toward an outcome. Providing feedback is vitally important.

Thus, if all of the motivational efforts fail with them, what is left? Actually, you have nothing unless there are consequences for their failure to be motivated by the previous efforts, for example, warning a worker to wear protective eyewear and even giving the worker a written warning. If nothing is done the next time he/she fails to comply with the safety rule, then the worker has no consequence. The worker's negative safety performance has been reinforced. But if the worker is given 3 days off without pay or dismissed, this says to others that a value on wearing safety eyewear exists. With that value comes a consequence for this unsafe behavior. There is a strong message that the company means business related to this safety rule. For without consequence, many individuals will not be motivated to perform as desired. Although a very negative approach, without it, a very critical motivational key is lost. It is possible for consequences to be positive. For example, a commendation for safe performance is a positive consequence.

If the previous four keys have not been successful, then the fifth component of using consequences must be put to use, followed by the sixth—reinforcement, seventh—involvement, eighth—self-monitoring, and ninth—rewards.

Reinforcement is an important key in the motivational process. Reinforcement of safety performance will determine whether it is strengthened or weakened. Reinforcers can be verbal feedback, a reward, or a consequence. It depends upon how the reinforcement is being used to drive home the message of accomplishment or failure to reach the safety and health goal. Telling a worker that the company really appreciates the safe way he/she is performing his/her job is feedback that reinforces the type of behavior that is desired and fosters motivation in that individual. Reinforcement for safety and health needs to be more frequent than once a year. Monthly would be best, but quarterly is also adequate. Unless the reward is very large, a year is too long a period to have to wait for reinforcement.

A key that has been discussed earlier is involvement. We should make every effort to involve workers who have a vested interest and invest their energy toward the outcome of a safe and healthy workplace. There are many ways to involve workers in the safety and health initiative. These range from participation in a safety and health committee to conducting inspections. This involvement needs to be nurtured and recognized. When our contributions are supported and recognized, it has a positive impact upon behavior, and thus, workers are more motivated.

Many times, workers like to be able to monitor their own progress toward a goal or expectation. If a chalkboard exists, workers themselves can mark the board each time that they or a fellow worker has gone without an injury for an agreed-upon period of time. Even if somewhat inaccurate, workers need to sense that they are involved in the process of safety and health at their workplace. This can lead to more teamwork and more motivation in the workplace. Self-monitoring may not always be an option, but do not overlook its impact when it can be used.

One of the most debated keys to motivation in the safety arena is rewards. In all aspects of life, we tend to focus and perform better when rewards are involved. Rewards are not a quick fix to problems with a company's safety and health performance. Rewards are a complement to the safety and health initiative. If a company does not have all the nuts and bolts of a safety and health program in place, then rewards are not a replacement for failure to effectively manage the safety and health effort.

The organized approach to safety and health should address each of the previous nine keys. Within the previous paragraphs are found the keys to motivation, which should be an integral part of the OSH prevention program initiative.

Simplify Motivation

Motivational attempts should be real and meant for people to achieve their intended purpose. Generally, the golden rule, "Do unto others as you would have them do unto to you," is a good rule for guidance. This is effective in all situations. There are also some specifics that good leaders need to be cognizant of and apply to the motivation of safety and health:

- **Communications**—Always keep people informed of what is going on within their organization. They like to feel they can be trusted with information when it becomes available. Make expectations clear, and follow the old adage, "Say what you mean and mean what you say." Make time available to meet with people, and make that time unhurried, and without interruption. Actively listen to what they are saying to you. Get to know the people you are trying to motivate and find out what their goals and aspirations really are.

- **Involve people**—Allow people the flexibility of being involved in the decisions that directly affect them. This will increase their personal commitment and their feelings of having some control over what impacts them. Include individuals in goal setting; this increases their stake in accomplishing the established goals. Let them know what part they play in accomplishing these goals and how they can contribute. It is critical to get everyone involved.

- **Responding to others**—Frequently, provide feedback to others. Whether the comments are positive or negative, do not wait until a specific time or until something is finished. Feedback is most effective when people are informed of how they are doing immediately following their performance.

- **Support others**—Help others reach their goals by offering advice and guidance, and recognize and reward good performance. In the work environment, help others get the rewards they deserve, and make every effort to get a raise or a promotion for individuals who warrant it.

- **Demonstrate respect for others**—When meeting with people, don't disrupt a meeting by answering the telephone; their time is also valuable to them. Avoid canceling or scheduling meetings at the last minute; this is indicative of poor preparation and the lack, again, of consideration for others' schedules. Do not reprimand another person in front of siblings, peers, or fellow workers; this will certainly result in ill will.

- **Be a role model**—No matter what the sacrifice that must be made, a leader cannot be an effective motivator unless he/she demonstrates and lives what he/she expects of others. A leader needs to take the lead by being prompt, conscientious, and consistent if others are expected to mimic leadership.

The Key Person

In maintaining communications, fostering good morale, attaining production goals, and assuring that workers are working safely, no other person is as important as the first-line supervisor. All that affects workers comes directly from the frontline supervisor. This includes all training, job communications, enforcement of safety and health rules, the company line, and feedback on the overall function of the company. Thus, the supervisor sets the tone for the motivational environment and is the role model upon which workers base their own degree of motivation. The supervisor's role is critical to the function of the jobsite and the efficient and effective accomplishment of all work activities. No other person has more control over the workforce than this individual. If the supervisor emphasizes production above safety and health, then his/her workforce will tend to be motivated in that direction. This is the reason that the supervisor's skills as a facilitator of people are more critical than the expertise related to the work being performed. These individuals need more training and support than anyone else on the jobsite.

Summary

It is evident that we spend a large portion of our lives either motivating ourselves or trying to motivate someone else. Thus, hopefully this part has given some insights on why motivation is an important component of OSH. In summary, some of the key traits critical to understanding how to motivate people are as follows:

1. People are self-motivated.
2. What people do seems logical and rational to them.
3. People are influenced by what is expected of them.
4. To each individual, the most important person is oneself.
5. People support what they create or are involved in.
6. Conflict is natural (normal) and can be used positively.
7. People prefer to keep things the way they are rather than make a change.
8. People are underutilized.

With these thoughts in mind, people can be motivated by the following:

- Allowing them involvement and participation
- Delegating responsibility with authority to them
- Effectively communicating with them
- Demonstrating concern and assisting them with counseling and coaching
- Being a good role model to them
- Having high expectations of them
- Providing rewards and promotions based upon their achievements

Workers need to know what is expected of them, what happens if they do not perform, and what their rewards, outcomes, or consequences will be. We are motivated by what we think the consequences of our actions will be. Those consequences should be immediate, certain, and positive if we expect them to be motivational.

Practically all our motivational attempts are geared toward peers or employees, and this is accomplished through their employers, fellow workers, or supervisors. These motivators are, among other things, a funnel, which directs all materials and information to those who need to be motivated. The motivator also directs or carries out the vast majority of learning.

People have many abilities and talents that they are unaware of or just don't use. Motivators are responsible for bringing out those hidden abilities and talents and channeling them toward the goals, outcomes, behaviors, and objectives desired. If this is done as discussed previously, it will give people a new sense of enthusiasm and self-esteem.

Motivation takes a lot of nurturing and caring for both the people involved to benefit and for the goals to be attained. Many organizations say that people are their most important asset but fail to exhibit that principle by the manner in which they treat their employees.

Motivation is not something that can be scheduled for a Thursday at 2:00 p.m. It is a process that requires continued commitment and a leader's ability to have an effect on others.

The essence of motivation is to find meaning in what an individual is doing. Motivation is the predisposition of doing something in order to satisfy a need. In real life, most people rarely have just one need; they have several needs at any one given time and are, consequently, moved to do something about them. Unfortunately, if they have too many needs facing them at one time, they may become indecisive, highly aggressive, negative, or even irrational.

Motivation is internal and can be stimulated by leadership and incentives. But, unless you know something about the needs, desires, and drives of the other person, your leadership and incentives may be completely ineffective. When people's task or job does not permit them to satisfy their own personal needs, they are less likely to work as hard at accomplishing the task that has been chosen for them. It seems safe to say that people do things well if they are excited about their assigned tasks. When their external environment assures that their own needs, wants, and desires will be met, it further enhances their desire to do a good job. As a leader/motivator, one has the responsibility of helping others meet the demands of their world according to the level of their capabilities. When each of these aspects is being fulfilled, an excellent motivational situation can transpire.

Further Readings

Blake, R.R. and J. Srygley. "Principles of Behavior for Sound Management," *Training and Development Journal* (October, 1979): pp. 26–28.

Blanchard, K. "How To Get Better Feedback," *Success* (June, 1991): p. 6.

Brown, P.L. and R.J. Presbie. *Behavior Modification in Business, Industry and Government*. Paltz, NY: Behavior Improvement Associates, 1976.

Frymier, J.R. *The Nature of Educational Method*. Charles E. Merrill Books, Inc., Columbus, OH, 1968.

Herzberg, F. "One More Time: How Do You Motivate Employees?" *Harvard Business Review* (January–February, 1968): pp. 53–62.

Maslow, A.H. *Motivation and Personality*. New York: Harper and Brothers, 1954.

Reese, C.D. *Accident/Incident Prevention Techniques*. New York: Taylor & Francis, 2001.

Reese, C.D. and J.V. Eidson. *Handbook of OSHA Construction Safety & Health (Second Edition)*. Boca Raton, FL: CRC/Lewis Publishers, 1999.

Weisinger, H. and N. Lobsewz. *Nobody's Perfect: How to Give Criticism and Get Results*. New York: Stratford Press, 1981.

25

Behavior-Based Safety

Behaviors are the deep-seated manifestation of a person's values, culture, and attitudes. With this in mind, safety and health professionals, managers, and businesses/companies have come to realize that their safety and health programs are not just an organizational effort. It has been shown that a total approach has to be undertaken with regard to individual differences in the actions of workers themselves.

Efforts have been made to understand why individuals perform at-risk acts such as the following:

- Failure to warn coworkers or to secure equipment
- Ignoring equipment/tool defects
- Improper lifting
- Improper working position
- Improper use of equipment:
  - At excessive speeds
  - Using defective equipment
  - Servicing moving equipment
- Operating equipment without authority
- Horseplay
- Making safety devices inoperable
- Drug misuse
- Alcohol misuse
- Violation of safety and health rules
- Failure to wear assigned personal protective equipment (PPE)

It has been suggested that 75% to 95% of accidents or incidents are due to risky behavior. Whether accurate or not, if these statistics are anywhere near accurate, then behavior-based safety (BBS) has a role in occupational safety and health (OSH).

Behavior-Based Safety Today

BBS cannot be viewed as the panacea or end-all solution for the prevention of accidents and incidents but only as one tool in the arsenal of tools, and it does not supplant a complete and organized overall approach in addressing OSH issues of today. All the components must be in place, such as training, a safety and health program, accident/incident analysis, safety engineering, controls, interventions, etc. It is only then that BBS can become an integral part of the whole OSH initiative.

It has always been a goal to get all employees to be motivated to perform their tasks in a safe and healthy manner, but this goal is only achievable when all safeguards are in place, all feasible protections are provided, all hazards are eliminated or controlled, safety and health is managed effectively, and the workplace has been structured to protect the workforce as best as possible, etc.; then, the application of behavioral approaches can be implemented to elicit a changed behavioral pattern and attitudes toward the standard practice of all the workforce. Individuals and groups take responsibility and become involved with the prevention process by adhering to policies, safe procedures/practices, and rules with regard to working and performance of all aspects of their jobs in a safe manner without any thought of circumvention of standards of practice or best practices.

This chapter is an explanation of what BBS is, the pros and cons of using a BBS approach, and why it is used. Definitely BBS is not a cure-all for safety and health issues, nor is it meant to replace the existing safety and health program. It often is viewed as a quick fix for safety and health problems but is not easy to implement. It is an organized approach and not common sense or a new way to blame employees.

Every BBS approach must be designed to fit the needs and culture of the organization or business. It is based upon the notion that safety and health is a shared responsibility and not just a personal matter. It is a way in which the employers provide the tools to optimize safety performance in the employees' unique work environment by developing methods to measure successes regarding safety performance in accomplishments, rather than using the traditional failure rates.

BBS Described

BBS is a process used to identify at-risk behaviors that are likely to cause injury to workers and is dependent upon the involvement of workers in this process so that they become willing participants and buy into the concepts and purpose of BBS. They will be asked to observe each other and their coworkers in order to determine if decreases of at-risk behavior have resulted in a reduction in these unsafe behaviors. This is a very simplistic description of BBS. Although there are an infinite number of variations to BBS programs, they all share common characteristics.

These are as follows:

- Identifying critical at-risk behaviors
- Gathering data and information
- Encouraging two-way feedback
- Stimulating continuous improvement

The most common barriers to safe performance must be addressed, or this is why a BBS program will fail. Some of these barriers are as follows:

- Hazard recognition: If workers did not realize that they were performing an at-risk behavior, then they could never perform the task in a safe manner.

- Business systems: The at-risk behavior was the result of an organization system that was unreliable due to inefficiency. If this occurs, workers will avoid using the system; they will find way around it.

- Disagreement on safe practices: There can be legitimate disagreement as to what constitutes safe performance, and this needs to be reviewed and addressed in some manner. This is best accomplished by working toward an agreed-upon consensus.

- Culture: The way that it was always done may be at odds with what is a safe practice. It is hard to teach old dogs new tricks.

- Inappropriate rewards: Rewards for achieving production may be at odds with safety and reinforce that at-risk shortcuts are more of a benefit than safe performance.

- Facilities and equipment: Outdated facilities or processes and rigged, missing, or damaged equipment may cause workers to act in an unsafe manner.

- Personal factors: This is when the at-risk behavior stems from personal characteristics of the worker that lead to him/her deliberately taking risks or refusing to work safely as a result of factors such as fatigue, medication, stress, or illness.

- Personal choice: A worker with adequate skill, knowledge, and resources chooses to work at risk to save time, effort, or something similar.

A successful approach is able to remove these barriers by observing and talking with employees and must not, in any manner, imply that the workers are the problem. It is important that the workers are viewed as the solution.

The idea that consequences control behavior is the foundation conceptually to BBS. Thus, the majority of behaviors rely on applying previous experience of consequences (both negative and positive) as the reinforcing factor. A picture of an amputated finger portrays visually the consequences of at-risk behavior, and this reminder of a negative consequence may be enough to cause a worker to alter his/her behavior prior to a similar incident.

Even though the previous barriers are addressed, there are no guarantees that BBS will work in your situation, but the principles are applicable to any situation when designed and implemented to meet businesses' or companies' needs.

Hindrances to Implementing BBS

It is never easy to implement anything new, and BBS is no different. This is why if the hindrances to implementing a program such as BBS cannot be overcome, then failure is guaranteed. Some of these deterrents are as follows:

- Failure to insure that all those expected to participate know about the basic principles and philosophies behind the program and how they can help and participate in its success; otherwise, it becomes just the newest kid on the block or, as some call it, the flavor of the month. To be fully successful, everyone needs to be a participant in order for BBS to become part of the company's culture.

- If the participants understand the principles and honest intent of the BBS program, then they will feel more at ease in taking ownership. If they do not have ownership, then they will not incorporate it into the culture of the company or organization.

- When it is realized that the workers performing the tasks are the true safety experts, they must be enticed to share responsibility for the implementation of BBS. Without support of the front-line supervisors, workers' failure is assured. Techniques to elicit bottom-up involvement are critical.

- Resources and lip service of BBS are important since no program functions without top-down support, but the real commitment is the physical presence from the top since what management does speaks louder than their words. Physical/visible presence is more powerful than other forms of support.

- Every program needs its cheerleaders. This means that management, supervisors, and labor leaders need to support BBS by being role models and extending themselves to commit time and effort. In order to be a leader for BBS, the most valued gift is to give of yourself and your time. A real champion is committed. Management must give these individuals the education and training as well as resource support in time and materials to facilitate the program.

- The mission of the organization or company is the high ground and usually very lofty, such as an accident-free or injury-free workplace, while the goal may be a percent reduction of accidents or injuries over a given period, which is a possible, realistic, and attainable goal. Progress toward this goal is measurable, and accountability can be obtained if the individual or group has direct control and resources to achieve it. Remember that purpose and goals are separate animals.

- It is difficult to effectively measure the outcomes of BBS since the reduction of injury rates can be manipulated and may not be a true indicator of the effectiveness of this program. Outcomes include the completion of observations, the number of individuals actually participating with deeds, the refining of at-risk behavior identification as denoted by extinction of some behavior, and the reporting of more at-risk behaviors, near misses, or first-aid cases. Measures that indicate changes in culture and successes in the program may be slow to evolve but are better positive reinforcers than just counting bodies.

This is why by addressing these potential hindrances to BBS, the company or organization can take steps toward an effective implementation of a successful program.

Summary

First, BBS is based on the general principle that behavior causes the majority of accidents, but this does not excuse employers from providing a safely engineered workplace with all controls in place to prevent the occurrence of incidents. Second, accountability inspires behavior, and accountability facilitates accomplishments. Third, feedback that fosters good communications is the key to continuous improvement, and excellence in safety needs to be established as the underlying culture desired in the organization or company.

These premises are driven by the following strategies:

- Obtaining of objective evidence of at-risk behaviors
- Defining barriers to safe behavior
- Teaching ways to substitute safe behavior for at-risk practices
- Holding employees accountable for improving their safety-related behaviors and helping others do the same
- Demonstrating the effectiveness by measurement that garners continued management support

BBS is used to increase safety awareness and to decrease accidents/incidents, by focusing on identification and elimination of unsafe behaviors. Workers are trained to conduct safety observation and give guidance on specific behaviors while collecting the information in a readily available format for providing immediate feedback. Observations are structured to have minimal impact on the workload, and data are shared with the entire workforce.

For such an approach to be successful, a good organizational safety culture and people's participation and involvement are required. Since real-time safety analysis is an integral part of BBS using operational personnel involvement to identify hazards and risks that are the key to effective BBS, the behavior-based process allows an organization or company to create and maintain a positive safety culture that continually reinforces safe behaviors over unsafe behaviors, which ultimately results in a reduction of risk.

The organization or company's purpose must be to continuously improve with the ultimate goal being a workplace that is free of injuries and illnesses. While attitudes are not addressed directly, it is a deep-seated intention to have employees accept safety as a value over time.

Further Readings

Federal Aviation Administration, *System Safety Handbook, Chapter 3*. Washington, DC: FAA, 2000.

Geller, E.S. *The Psychology of Safety Handbook*. Boca Raton, FL: CRC Press/Lewis Publishers, 2001.

Heinrich, H.W. *Industrial Accident Prevention: A Scientific Approach*. New York: McGraw-Hill, 1959.

Swartz, G. (Editor). *Safety Culture and Effective Safety Management*. Itasca, IL: National Safety Council, 2000.

Wieneke, R.E., J.J. Balkey, and J.F. Kleinsteuber. *Success in Behavior-Based Safety at Los Alamos National Laboratory's Plutonium Facility*. Los Alamos, NM: Los Alamos National Laboratory, 2002.

26

Safety Culture

It is not at all unusual that a company or business develops a comprehensive occupational safety and health program (OSH) and then sets about implementing it in their various geographic locations and facilities. In some locations, it is very successful, while in others, it is a total failure. Why is this possible? Because the organization or company has failed to consider that the safety culture in each place of implementation is not the same. Suffice it to say, it is imperative that the culture of managers and employees is unique in these varied facilities.

Defining Safety Culture

Safety culture is a concept defined at a group level or higher that refers to the shared values among all the group or company, corporation, or organization members. Safety culture is concerned with formal safety issues in an organization and is closely related to, but not restricted to, the management and supervisory system. Safety culture emphasizes the contribution from everyone at every level of the organization. The safety culture of the business entity has an impact on all members of the workforce's behavior at work. Safety culture is usually reflected in the relationship between the reward system and safety performance. A positive safety culture is indicative of an organization's willingness to develop, change, and learn from errors, incidents, and accidents. Safety culture is ingrained, enduring, stable, and very resistant to change.

In summary, safety culture is the enduring values and priority placed on workers and management by everyone within the organization. It refers to the extent to which individuals and groups will commit to personal responsibility for safety; act to preserve, enhance, and communicate safety concerns; strive to actively learn, adapt, and modify behavior (of both individuals and the organization) based on lessons learned from mistakes; and be rewarded in a manner consistent with these values.

Developing or Changing a Safety Culture

Without exception, development or change starts with management's commitment by actions, not words. Words have no substance, but action or behaviors have substance. During this process, actions and development must be maintained by a continuous, constant, and consistent effort. It must be worked at, a new plan must exist, a new approach must be fostered, and support must be gained from all involved.

During development and change, individuals are not merely encouraged to work toward change: they must take action when it is needed. Inaction in the face of safety problems must be unacceptable. This will result in pressure from all facets, both peers and leaders, to achieve a change of culture. There is no room in a culture of safety for those who uselessly point fingers and say, "Safety is not my responsibility" and file a report and turn their back on helping to fix it.

Senior management must support and commit to culture change by demonstrating commitment to safety by providing resources to achieve results. The message must be consistent and sustained since it takes a long time to develop or change a safety culture.

The workplace culture is not something that can be categorized into certain specific types of culture. Culture is what everyone in the workplace believes about the company, themselves, and safety. These opinions, assumptions, values, perceptions, stereotypes, rituals, leadership, and stories all mesh together to form the culture, which translates into policies, procedures, and accident/incidents. There are many factors invisible from the surface, the taboos, assumptions, and norms, which are never written down. These are the true forces behind outward safety behavior that reflects the real value of safety within the organization. No one espouses these deeply buried parts of the culture, but everyone knows what they are.

Many factors impact the culture of the workplace. They include the following:

- Intelligence and job knowledge
- Emotions and emotional illness
- Individual motivation toward safe work
- Physical characteristics and handicaps
- Family situations
- Peer groups
- The company itself
- Existing society and its values
- Consequence of the work itself

Since everyone at the workplace knows what the culture is, it is necessary to be very observant and listen prior to trying to communicate. It does not take long to discover the culture. It is seen in the safety, productivity, quality, and discipline of the work and the workforce. The culture often stymies efforts to communicate even when the outcomes are good. It is the sovereign duty of all to maintain the old culture as it is. Thus, the process of changing the culture is time consuming and quite complex.

It is bad enough that the workforce has its own culture, but this is compounded by the company's culture, which impacts, and at times is diametrically opposed to, the culture of the workforce. The goal would be to meld the two cultures together in a viable and productive relationship.

Industries and companies that have labor unions seldom are able to perfectly integrate safety into their culture even though safety should be in everyone's best interest. Unions often hold safety hostage as a bargaining chip, tool, or weapon against management. Thus, the culture does not see safety as a level playing field on which everyone can participate equally. But in the same light, management often fights the battles related to safety and health regulations, which were instituted to provide safety and health for its workforce, and is intent on fighting compliance.

Thus, the culture that exists is an amalgamation of the culture of labor and of the company. At times, there is reasonable compatibility, and the two parts form a culture that strives for effective safety, while at other times, they are so opposite that no one benefits from their inability to merge the culture into a useful entity where effective and rational communications occur.

The only way the culture can be changed is when an urgent need to change is the motivator. This could be a number of occupational deaths, a catastrophe, or a series of occupational injuries or illnesses, which forge partnerships in troubled times. Second, the resources must be available, and the ability to change needs to be present. Thirdly, a road map or plan must be developed and agreed upon by all parties involved in order to transition to the new culture. Most of us are reticent to change. Thus, gaining consensus for change is going to be difficult. It will be the leadership's responsibility to decide that there is a need for change, communicating it to the organization, getting consensus of all parties, and directing the implementation.

If this discussion of culture seems ominous and overwhelming, then you have some feeling of how difficult it is to communicate if the culture is not receptive to the message. This makes it imperative to understand the culture that one is trying to communicate within. As for safety, if the culture is antisafety, then the communications regarding safety will fall on deaf ears. So in order to communicate safety in the workplace, it may first be necessary to understand the culture and then begin to change it. This is a communicate-or-lose situation, which is not the best of all worlds. Suffice it to say that in most cases, the improved safety communications goal will be accomplished to some degree, based in a large part on the perception of how one really feels about safety and health at any workplace.

Positive Safety Culture

An organization, corporation, or company with a positive safety culture is one that gives appropriate priority to safety and realizes that OSH has to be managed like other areas of the business. Safety culture is more than just avoiding accidents or even reducing the number of accidents, although these are most likely to be the outcomes or measurement of a positive or successful safety culture.

A positive safety culture is doing the right thing at the right time in response to normal or emergency operation. The quality and effectiveness of training will play an important role in determining the attitudes and performance toward safe production or performance. These attitudes toward safety are in large part a mirror image of the culture set up by the company or corporation.

The keys to achieving a positive safety culture are as follows:

- Recognizing that all accidents are preventable by following correct procedures and established best practices
- Maintaining constant awareness and thinking about safety
- Trying to improve safety on a continuous basis

The goal is to have effectively trained workers who are motivated to self-regulate and take personal responsibility to work safely using the safest practices where these have become an integrated part of the workers' value system and company culture.

Assessing Safety Culture

Answering the following questions regarding the safety culture at a place of business or operation will help determine the effectiveness of safety culture.

- Is safety and health a top priority in the organization?
- Is safe work performance reinforced, recognized, and rewarded?
- Are business decisions made with safety and health being given major consideration?
- Is the requirement that safe and healthy attitudes and behaviors are expected of all employees including management?
- Do employees feel comfortable voicing their concerns regarding any hazards or safety issues about the workplace to management?
- Does peer pressure act as a positive or negative in support of a positive safety culture?
- Do production deadlines cause safe and healthy work practices to be overlooked?

Summary

If the safety culture does not coincide or integrate well with the business approach of the organization, corporation, or company, the indication is that a safety culture does not exist or only marginally exists. A positive safety culture exhibits the following characteristics:

- The importance of committed leadership from top to bottom.
- A set of clearly explained expectations for line management.
- The involvement of all employees.
- Effective communication must exist.
- A commonly understood and agreed-upon set of safety and health goals.
- A learning organization that is responsive to change.
- A zeal and attention to detail regarding safety and health.

Further Readings

Reese, C.D. *Accident/Incident Prevention Techniques (Second Edition)*. Boca Raton, FL: CRC Press, 2012.

Reese, C.D. *Occupational Health and Safety Management (Third Edition)*. Boca Raton, FL: CRC Press, 2016.

Zhang, H., D.A. Wiegmann, T.L. von Thaden, G. Sharma, and A.A. Mitchell. Safety Culture: A Concept in Chaos? *Proceedings of the 46th Annual Meeting of the Human Factors and Ergonomics Society*. Santa Monica, 2002.

27

Communicating Safety and Health

Effective communications is a lot more complex than most think it is. In fact, most individuals feel that they do a good job of communicating when in actuality, most do a rather ineffective job of communicating safety and health in the workplace. Sometimes it is the way the message is communicated. The timing of messages is often too late, and in many cases, messages are never delivered at all. No matter how hard someone tries, someone else is going to misinterpret the message, read something into it other than what was meant, or only focus on the part he/she perceives to be most useful to him/her.

Finding ways to communicate safety and health messages can be more of a task if your supervisors and workforce do not perceive its importance or meaning to them. In addition, their perception of the amount of real risk comes into play. If the company or business has a rather safe operation, almost everyone feels that there is no need to emphasize occupational safety and health (OSH) since little risk exists. They may make statements such as, "The company has only had one accident in the last 2 years." "It could have happened to anyone." "It was just bad luck. But when John fell, he broke his thighbone. The doctors put a pin and screws in it. He'll be as good as new." John did not return to work for 6 months.

What has not been communicated to the workforce is that John will never be as good as new. It took him 6 months to return to work, and workers' compensation did not pay him his full and normal wages during that time. John's medical bills cost $50,000, and the company's insurance premiums increased because of this. Someone was hired who was not as skilled as John to do his job, which could create a further hazard. John was a friend to everyone. He knew better than to jump off a piece of equipment. He was an experienced worker, but he disobeyed the safety and health rules.

This scenario is not unusual. Similar situations are played out everyday across the 6 million workplaces in the United States. How can a company address its safety and health concerns? It definitely starts with management's commitment and attitude toward OSH. If the motivational fire or the drive to push safety and health in the workplace is not there, then the communications attempts will be viewed as nothing more than a noise from a hollow log. There is a saying that someone cannot give the measles to someone else unless he/she has had the measles. This same logic goes for communicating the company's message on safety and health. A company must truly believe that it is important to foster effective communications in the workplace.

Many safety problems arise because we assume that everyone knows the proper and safe way to do a job. In actual practice, this is not so. It is imperative that management ensures that everyone on the property knows the safety and policies of the company and the proper methods to use in performing their job. This comes about by effective training and good communications.

A large percentage of injuries and illnesses occur when people are not aware of the policies, methods, or basic skills needed to perform a job safely. The responsibility for communicating these concepts rests with the management but is often assigned to the supervisor, who is already overloaded and does not see safety and health communications as an important part of his/her job. After all, there are no consequences to him/her if his/her communications are not effective.

A line of communication that constantly furnishes information to all employees must be established. Some methods of communicating the safety and health message are as follows:

- Management setting the example by abiding by their own rules
- Safety meetings
- Job training
- Joint/labor management committees
- Employee safety representatives
- Employee involvement
- Safety bulletin boards
- Use of computers (e-mails)
- Use of electronics signs

All of these must be forthright, sincere, honest, and consistently utilized in order for them to be more than lip service. What a company does speaks so loud that the employee cannot hear what they say. It is important to walk the talk.

The Communicator

A failure to communicate effectively is planning to fail. What this means is that the company and its management are the primary conveyors of OSH messages, whether it be a manager, supervisor, or safety and health professional. To paraphrase Ralph Waldo Emerson, how you act speaks so loudly, no one can hear what you say. If the company representative fails to wear the required personal protective equipment in designated areas, then his/her actions don't match the message the company wants to send. This is a communication failure, plain and simple.

This means that the failure to visibly support occupational safety and health by their actions, follow the company's safety and health rules, or wear the required personal protective equipment sends a stronger nonverbal message than words ever can and will undermine the safety and health at the work site. The message communicator, whether he/she likes it or not, must be a role model for OSH messaging. Effective communication goes well beyond the written and verbal forms of communication. Communicators must walk the walk before they can credibly talk the talk.

This does not suggest that the company should not use written communications or vocally espouse support for safety and health at the workplace. However, this is all for naught when the safety and health rules and policies are sidestepped when it is inconvenient to follow them. When this occurs, it is perceived that the company is placing production above safety and health in the workplace.

Is this what is meant to be communicated? If the company looks the other way when supervisors do not follow the OSH policies and procedures, the company's actions are more effective than anything that has been said or written regarding commitment to safety and health.

Before attempting to set up communications that will work in the company's unique situation, the company needs to make sure that they have a good understanding of the culture that exists, and this is especially true of the safety and health culture. The safety and health culture needs to be assessed to make sure that the communications will be received in a manner that will positively influence the safety and health effort.

Communication Tools

The following are some the communication techniques that can be used to make communications related to safety and health more effective.

- Written materials (paycheck stuffers)
- Bulletin boards
- Electronic signs
- Computers
- Posters
- Public address systems
- Safety and health talks

Written Materials

Written materials should be ergonomically sound so that they are easily readable, legible, and understandable by the workforce. They should also be attention getting. This can be accomplished by using black print on white or yellow paper, as this is more visually favorable than other combinations. Make the message as simple as possible. Do not make the page(s) too cluttered. Actually, double columns (such as those used by newspapers) are more readable than single ones.

It is a wasted communication tool if no one will read it. Your workforce may have poor reading skills, or the culture of the workforce may be such that they do not like to read. If this is the case, you are wasting your money investing in written type of materials.

Written materials are only useful for a short time period. It is a good idea to use them sparingly, maybe once a month as a stuffer in the payroll envelope or during a safety talk. If you post it on the bulletin board, it will be read within a week. You should then take it down since it has probably been read by all of those who will take the time to read it. At times, you may develop plasticized cards that contain new rules, inspection directions, operator guidelines, or changes in procedures, which you cannot expect a worker to memorize. These cards should fit into a shirt pocket and can act as reminders to workers if they have questions when no one is around. At times, it is recommended that safe operating procedures be plasticized and placed on machinery or equipment that is complicated to operate or not operated frequently. If you use written materials, test them out on the audience before mass distribution. Written materials have their place in fostering workplace safety and health communications.

Bulletin Boards

Well-designed and maintained bulletin boards can communicate safety and health messages. The most effective bulletin boards are those placed in well-lit areas convenient to the workforce. They are usually spacious and well designed to provide good spacing between items. The bulletin board should have a single focus on safety and health specifically to display information, for example, on accident statistics or confined spaces. It must be kept current especially if statistical data are displayed. In other words, a 6-month-old date should not be displayed if monthly data are available. You should change the topic on a monthly basis to keep it fresh, i.e., switch from lifting to ladder safety.

Electronic Signs

A computer-controlled electronic message sign in the workplace, along highways, and outside of businesses is an important communication tool. The advantage is that the message can be changed and updated quickly. The message, alert, or notice is, or can be, a function of real time. It can be easily changed to convey the new hot topic. This also compels workers to pay attention to it so that they do not miss anything that is new or important to them. One company used an electronic sign thanking the workers for completing a special order without any injuries. These types of signs may be one of the best safety and health communication tools. If you have a large facility with multiple entrances, you will need to have one for each entrance. If these signs help keep current safety and health issues on the workers' minds, then they are probably well worth the cost.

Computers

The computer appears in almost all workplaces today, and a large part of the workforce has a computer at their workstation or at least ready access to one. It behooves businesses and companies to make use of this very fast communication device. It can be used for alerting workers to emergencies, timely reporting of potentially harmful hazards, short safety alerts, and safety messaging using instant message capabilities, or for providing short or long training modules that can be accessed at the workers' workstation or at another time, such as at home on their own computer. Instantaneous feedback can be provided on training modules regarding the success of completion or feedback.

The computer can target specific workers who may need special attention regarding safety and health issues. It can be used to post short safety messages to individual workers or work groups, or address family safety issues. It can be used to submit safety suggestions and respond to questions posed regarding safety and health issues. It can readily disperse time-sensitive information on safety and health organization-wide at the push of a button. The only limitation is the creativity of the operator. The computer is probably the number one communications tool.

Safety and Health Posters

The posters seen in many workplaces that communicate a number of messages relevant to safety and health with color and graphics are eye-catching and provide a simplistic safety and health message. These posters usually have one message at a time. You must have a poster program to ensure that you are constantly rotating or replacing posters on at least a monthly basis. The posters become stale with time. They might impress visitors, but they quickly lose effectiveness with the workforce if they are not current and are ever changing.

Public Address System

If your company has a public address system that reaches all areas of the workplace and loud noise is not a problem, then short safety and health messages or alerts can be communicated during the workday. These types of announcements should be written ahead of time and should not be announced so often that they become something the workers tune out. They could be a reminder of a hazard that exists, an explanation of an accident/incident, or the safety message of the day. These messages should be short and to the point.

Safety Talks

Safety talks are especially important to supervisors in the workplace and on work sites because they afford each supervisor the opportunity to convey, in a timely manner, important information to workers. Safety talks may not be as effective as one-on-one communications, but they still surpass a memorandum or written message. In the 5 to 10 minutes before the workday, during a shift, at a break, or as needed, this technique helps communicate time-sensitive information to a department, crew, or work team.

Summary

Plan your lines of communication and keep them open. A safety and health program or effort may be good on paper, but unless it is communicated to the workers, it is useless. Even more important is the company's credibility. Does the company mean what they say about commitment to safety and health? Do they demonstrate that commitment? Is the organization always reinforcing the safety performance that they desire? Continue to work on communicating. Failure to communicate is like the man who winks in the dark and no one knows it. Communications regarding safety and health are vital to an efficient safety and health program.

Further Readings

Reese, C.D. *Accident/Incident Prevention Techniques (Second Edition)*. Boca Raton, FL: CRC Press, 2012.

Reese, C.D. *Occupational Health and Safety Management (Third Edition)*. Boca Raton, FL: CRC Press, 2016.

Reese, C.D. and J.V. Eidson. *Handbook of OSHA Construction Safety & Health (Second Edition)*. Boca Raton, FL: CRC/Lewis Publishers, 2006.

Zhang, H., D.A. Wiegmann, T.L. von Thaden, G. Sharma, and A.A. Mitchell. Safety Culture: A Concept in Chaos? *Proceedings of the 46th Annual Meeting of the Human Factors and Ergonomics Society*. Santa Monica, 2002.

28

Bullying

Bullying has always existed but has not been addressed as an occupational safety and health (OSH) issue until recently. It has become more of a concern with the rise of workplace violence as an OSH issue. Workplace bullying is not always thought of as an aspect of safety and health. It is commonly viewed as more of a social problem that is interpersonal in nature, one that individuals should work out between themselves. However, it is not that simple. The reasons for addressing bullying are that the workers have a right to expect to not be faced with a hostile work environment or experience adversarial relationships within their place of employment. Bullying has been linked to mental and physical illness in the workforce. The impact of this type of dissension does not impact favorably on productivity.

Suffice it to say that the reasons bullying impacts work are as follows:

- Makes the workplace uncomfortable and hostile.
- Decreases or makes the workplace less than ideal.
- Makes the work environment intimidating.
- Tolerating bullying sends a negative message related to management's stance.
- Workers feel less safe.
- May result in workplace mental illness as well as physical illness.

Some recent studies indicate that individuals being bullied have a higher risk of heart attacks and other stress-related illnesses. Bullying may not be the definitive cause but is more than likely a contributory fact in those experiencing what they perceive as being bullied.

A definition of bullying (psychological harassment or emotional abuse) is as follows: the result of the conscious repeated effort to mentally wound and seriously harm another person with words and actions, not violence. The bully attempts to steal the target's self-confidence. Bullies are most often bosses, managers, supervisors, and executives. It is interesting to note that 35% of workers experience bullying personally, and in 62% of the cases the bully was male, while 34% of the targets were women, as determined by surveys in 2011 from the Workplace Bullying Institute (WBI).

Workplace bullying is a human phenomenon that is inherent in most settings where individuals must work together on a daily basis, as part of a team, factory, work facility, department, or office, or in the public setting. Myriad relationships develop in any workplace. They can range from formal to informal, brief to long term, and implied to intentional, and are social by nature. In those relationships, there are gender factors. Other defining aspects include age, race, ethnic background, and more. However, suspiciously, those aspects that are typically associated with discrimination are not necessarily the same ones that show up when bullying occurs. This is why bullying is viewed as negatively impacting the workplace.

One would think that bullying would be considered discrimination in the workplace. However, that is not the clear rule, nor do company policy, government regulation, and law define it as such in all cases. Approximately 80% of companies are doing nothing to address bullying or actually resist action when requested to do something. Its prevalence is more widely accepted than in the past, but the research and study of bullying have yet to result in the same level of importance as an issue to be prevented through government regulation and law. It is believed that the psychological aspect of this hazard is difficult to pin down, and in many cases, there is a hesitance to reach a unanimous conclusion to treat it, prevent it, or invoke punishment for its results on the livelihood of workers. In some ways, many of us may wonder if we have been the victim or target of bullying in the past. We may also wonder if we have been the bully in our actions.

Data Regarding Bullying

According to the Workplace Bullying and Trauma Institute (WBTI) and the research led by Dr. Gary Namie, there are many specifics that can be tallied from well-recognized, national surveys taken in 2000 and 2003 from workplace respondents who were targets of this particular workplace hazard. The total number of respondents was 2,335 individuals from all walks of life, making it the largest such survey of its kind done on this subject. The facts speak for themselves:

- Of the individuals targeted, in 2003, 80% were women and 20% were men, and in 2000, 77% were women and 23% were men.
- Average age of the targets was 43 (mean).
- Eighty-four percent were college educated, 63% degreed, 17% with graduate degrees, and 4% were PhDs, doctors, or lawyers.
- Length of career was a mean of 21.4 years, with a total of 6.7 years in the workplace where the bullying occurred.
- Type of employer was 36% corporate, 31% government, 12% nonprofit, and 11% small business.
- Of coworkers, 96% were aware of the target's abuse. The targets had told 87% of their coworkers.
- Average length of time that the target was exposed to bullying was 16.5 months, men reporting 18.38 months and women reporting 15.74 months.
- Majority (67%) reported that they had never been a victim before. They also were not victims of trauma in the past (62%).
- Bullies had targeted others 77% of the time on average and 88% of the time in government employment.

Of the targets, 37% were fired, 33% quit employment, and 17% transferred to another position that took them away from the bully. The effect on the individual target was dramatic. The Canadian Safety Council found that the targets of bullying wasted between 10% and 52% of their time at work consumed by the effects of their tormenter. They are demotivated, highly stressed, absent more often, fearful, anxious, and angered, and 41% of the time, they become depressed. It is a well-known statistic that depression affects about 10% of the workforce at any one time, and the estimated cost of untreated depression in the workplace is $24 billion annually in lost productivity alone. Depression results in 200 million lost workdays and costs the US economy $43.7 billion annually. According to the World Health Organization, depression may affect up to 20% of the world's workforce at a global cost of $240 billion. This means that the bully in the workplace has the potential to cause substantial economic damage to his/her own company through his/her actions. As can be seen from the data, this is why bullying should be considered a major problem in the workplace.

The effects go on to include the family of the targets, damaging marriages and transferring feelings of guilt and shame onto their children. Physical aches and pain are common, including headaches, nausea, stomachaches, skin rashes, sleep difficulties, and other stress-related illnesses. Stress can lead to heart disease, suicide, and diabetes, and complicate a host of other illnesses.

The bullies have their own set of statistics. The data derived from the same surveys indicate the following:

- Women were 58% of the bullies, while men were 42% of the bullies.
- In only 25% of the cases was the target a member of a protected class and the bully was not. In 15% of the cases, the bully was the protected class.
- Women bullies chose women targets 87% of the time and chose men 13% of the time. Men who were bullies chose women targets 71% of the time and chose men 29% of the time.
- Woman-on-woman bullying represents 50% of all workplace incidents. Man on woman was 30%, man on man was 12%, and woman on man was 8%.
- Probability of women targets to be bullied by women is 63%, and the probability for men to be the targets of men is 62%.
- Bullies had a higher rank than the target 71% of the time: for men, 76%, and for women, 67%.
- Only 17% of the time was the bully a coworker or peer of the target.
- In some cases, workers such as supervisors are bullied by their employees. This occurs to women 36% of the time, while men suffer this fate 23% of the time.
- Women were more likely to target more than one worker 68% of the time, while men would target more than one 63% of the time.
- Bullies did the targeting by themselves only 23% of the time; 77% of the time, they enlisted the help of others, by bullying another target to help 32% of the time and using some help from others 45% of the time.
- Targets' coworkers were the bully's help 48% of the time. Women enlisted help 53% of the time, and men, 42%.
- When bullies worked in groups, they were an average of 3.5 people: male groups, 3.9 people, and women, 3.2.

Bullies were overwhelmingly bosses who had the authority to terminate their targets at will. Bullies typically possess a type A personality; they are competitive and appear driven, and operate with a sense of urgency. Bullies tend to crave power and control. Bullies use charm and deceit to further their own agenda. Those who are working around a bully often seem to walk on eggshells since there is uncertainty as to what will set the sleeping giant off.

Facts about Bullying

What is a bully, and what are the typical characteristics? Here are some definitions from a wide sample of research that is attempting to address the problem. This is an excerpt of facts regarding bullying from the 2003 Report on Abusive Workplaces sponsored by the WBTI; the principal investigator was Gary Namie, PhD:

- Bullying is the repeated, health endangering mistreatment of a person (the target) by a cruel perpetrator, the bully.
- It is best understood by the bully's behaviors—acts of commission and omission—which are all driven by the bully's need to control other people.
- Initially involves the bully deciding who is targeted, when, and where and psychological violence inflicted. Later, others may be coerced to participate in the assault.
- Illegitimate behavior as opposed to tough behavior, it interferes with an employee's work production and the employer's business interests.
- It escalates from 1:1 harassment before being reported followed by a limited or nonexistent employer response, eventually engulfing an entire work unit in fear, paralyzing productivity.

The following are some of the characteristics that have been recognized by the same institute in a report titled US Hostile Workplace Survey, 2000. Dr. Gary Namie was the research director:

- Screaming, yelling, public attempts to humiliate, seeking to do battle when and where he or she chooses, needs to compete and win to feel good.
- Controls all resources (time, budget, support, and training) so as to prevent you from being successful at your job, undermining, set you up to fail.
- Constant, personal verbal assaults on your character, name calling, belittling, zealous attention to unimportant details, committed to systematic destruction of your confidence in your competence.
- Manipulates the impression others have of you, splits the work group into taking sides, and defames you with higher-up and at next job, killing your reputation.
- Bullies prefer public sites in front of witnesses for humiliating their targets. Occasionally the bullying was private and sometimes behind closed doors with the intent of being heard.

The following are the results of a study on bullying from the Mental Health Association of New South Wales, Australia, published by the Mental Health Information Service in conjunction with the Australian Council of Trade Unions. Workplace bullying is a serious safety and health hazard. It is usually characterized by the following:

- Unreasonable demands and impossible targets
- Restrictive and petty work rules
- Being required to perform tasks without adequate training

- Being forced to stay back to finish work or additional tasks
- Compulsory overtime, unfair work schedules or allocation of work
- Constant intrusive surveillance or monitoring
- Having no say in how your job is done
- Interference with personal belongings or sabotage of work
- Shouting or using abusive language
- Open or implied threat of termination or demotion
- Oppressive, unhappy work environment
- People being afraid to speak up about conditions, behaviors, or health and safety

Why Bullying Occurs?

First, we need to look at the individual who is the bully. Bullies share the same characteristics from person to person, although the individual people are quite different in many ways. The bully is typically punished only 8% of the time, while the bully's boss, if he/she is aware of the problem, will side with the bully 42% of the time. The bullying stops when the target goes away.

We know the basic characteristics as they mimic the general behaviors of disruptive employees, typically managers. They are rude, condescending with the intention to insult and demean their targets. Verbally, they sound hostile and angry. They may shout, throw things, slam doors, and berate staff in front of others. They seem to have an inherent insensitivity toward others' lives. They are often disrespectful to their peers, their support staff, and anyone they consider less important than themselves. They usually do not take criticism well and will counterattack anyone who directs blame at them. Bullies generally dislike authority and believe they are above the rules and policies that apply to others.

Psychologically, they are insecure about themselves and their place in the world. They may have chronic conditions that complicate their negative feelings toward themselves and others. There are cases of alcohol and substance abuse, family problems, broken marriages, and financial problems. These are all things that everyone has to deal with at some time in their lives, but bullies have not learned to cope and work through the personal issues that life's challenges present. Instead, they look to release frustration by targeting those that they see as vulnerable or deserving of some of the misery they themselves endure. As far as personalities are concerned, some are simply unpleasant people who have found that they can get what they want by imposing their wrath on others and do not care or understand the impact they leave behind.

In the worst-case scenario, two potential disasters can occur. Bullies can reach a point of frustration and emotional instability, and their actions are played out as physical violence directed at a target. The consequences now multiply tenfold. There is injury or even death to the target, the bully, or even innocent bystanders. Workplace morale will suffer its worst blow, with employees suffering psychological and physical health problems, poor productivity in the workplace, and the potential for company liability for creating conditions conducive to violence, or worse, not doing anything about obvious warning signs. The other potential disaster comes from the targeted employee. It happens where the emotions and psychological pressures brought on by bullying become too much to bear. The unstable person, who may be predisposed to a breakdown, suffering depression, or other psychotic illness, becomes the time bomb. We have heard about it in the news; it has happened before. This suffering individual decides that he/she must place blame on someone other than himself/herself and shows up at work with the intent to maim or kill the perceived tormenter and possibly others who may have played a part or been totally innocent and just bystanders.

Why Prevention of Bullying?

The prevention of workplace bullying must be approached like any other hazard, through a workplace and hazard analysis, research and compilation of the data collected, consultation with all appropriate management personnel, the hiring or contracting of expert services that are suited to recommend a plan of action, the development of the plan and its incorporation into the company's overall safety and health plan, and the assignment of roles and responsibilities. (Putting a plan into action includes the necessary training of management; the rollout of the plan, including the implementation of training and awareness of all employees; and finally, a method of measurement of the impact of the plan that will prove positive results. In the process, it may be necessary to address existing problems, as they are uncovered, through counseling, discipline, or other means.)

In establishing a plan and making it a part of corporate culture that is in line with business goals, you must be careful to use the appropriate resources to understand bullying and prevent it from becoming a workplace hazard. There are specific experts in a variety of fields who can be tapped for guidance. Certainly, a company safety and health officer should take the lead in the organization, and actions should be taken. A list of professionals who could be consulted would include industrial and occupational psychologists, occupational therapists, occupational physicians, occupational health nurses, risk management professionals, industrial security specialists, design or process engineers, organizational psychologists, and other organizational consultants.

Other specific tools can help analyze the workplace. The Occupational Safety and Health Administration (OSHA) has voluntary and advisory guidelines relating to workplace violence, stress reduction, conflict resolution, risk assessment, and health issues related to this hazard. The National Institute for Occupational Safety and Health (NIOSH) has similar resources available. Many other private organizations will provide information and leads to professionals who practice the services needed. Of course, all of these organizations should be used throughout the process of developing and implementing the plan.

The tools for understanding the problem of or potential for bullying, as a workplace hazard, are similar to those used to assess workplace violence, stress, and mental and physical health. They include the following: analysis of existing records; the monitoring of trends, incidents, and illness; employee and management surveys; focus groups; performance reviews; administrative and organizational procedure review; security and risk assessment; as well as outside observation of the business operation to lend an unbiased view of current activities.

One of the areas that managers spend considerable time on is conflict resolution. It may be as much as 20% of their time. Effective training in conflict resolution can help managers not only deal with bullies and their targets but also manage themselves and their own behavior. Additional training might include the general employee population on how to deal with conflict, change, and competition. Union representatives may also be able to provide training, advice, and mediation among employees they represent. Key points to bring into any of this training are the need for open communication, management commitment to address and prevent bullying, and a no-retaliation policy for those who bring any hazard to the attention of management. Most of all, convince people to bring bullying out into the open, and expose bullies for what they are and the damage that is done.

Management must take corrective action, uphold company policy, and administer disciplinary procedures up to and including termination. The rules and policies should already be in place to set and communicate expectations, measure performance against those expectations, provide feedback, manage poor performance with corrective actions, and document everything.

Summary

Bullying is a form of organizational violence and a potential source of work-related stress. Bullying is primarily an employee relations issue, best dealt with by employers' internal and disciplinary procedures before it becomes a risk to an employee's health. Bullying is usually viewed as occurring to schoolchildren but often happens in the work environment. As with children, adults feel threatened or ill at ease in reporting perceived incidents of workplace bullying. Probably many adult workers even fail to recognize that a behavior could be construed as bullying. Bullying is a behavior that should not be tolerated by employers, but the employer may need to educate all employees regarding what constitutes bullying and what employees are to do if they feel that bullying is taking place. Guidelines need to be developed to indicate to employees how to handle and address bullying.

This is why the company should have a policy and enforce it, as well as take bullying seriously, which has not been the case in the past. The company or organization departments that are most likely to be involved in bullying include senior management, human resources, and the safety and health point person for the company. Companies following the organizational approach had developed a set of values and convictions that reflected on the treatment of antisocial behavior and management response. A distinct element in these organizations was that they each had a reason for developing these plans that was the result of factors outside the company, such as a new state regulation or prior litigation that had cost the company time and expense. The leaders in these companies recognized that in order to be effective, they had to grow trust and support from the inside of their companies, which taught managers the importance of their roles in influencing the behavior of their employees.

Further Readings

Australian Health & Safety Commission, "Violence and Bullying at Work." http://www.work placeviolence911.com, December 2003.

Canadian Safety Council. "Bullying in the Workplace." http://www.safety-council.org, 2004.

Feiberg, P. "Bullying in the Workplace Is a Violence Warning Sign". http://www.workplaceviolence911 .com, July 1998.

Namie, G. "US Hostility Workplace Survey." The Workplace Bullying & Trauma Institute, September 2000.

Namie, G. "Report on Abusive Workplaces." The Workplace Bullying & Trauma Institute, October 2003.

Reese, C.D. *Occupational Health and Safety Management (Third Edition)*. Boca Raton, FL: CRC Press, 2016.

Yandrick, R.M. "Lurking in the Shadows." *HR Magazine 44:10*. October 1999.

29

Safety (Toolbox) Talks

Safety talks are especially important to supervisors in the workplace and on work sites because they afford each supervisor the opportunity to convey, in a timely manner, important information to workers. Safety talks, also known as toolbox talks, may not be as effective as one-on-one communications, but they still surpass a memorandum or written message. In the 5 to 10 minutes before the workday, during a shift, at a break, or as needed, this technique helps communicate time-sensitive information to a department, crew, or work team.

In these short succinct meetings, supervisors convey changes in work practices, short training modules, facts related to an accident or injury/illness, specific job instructions, policies and procedures, changes in rules and regulations, or other forms of information that the supervisor feels are important to every worker under his/her supervision.

Although safety talks are short, they should not become just a routine part of the workday. Thus, to be effective, they must cover current concerns or information, be relevant to the job, and have value to the workers. Plan safety talks carefully to effectively convey a specific message and a real incident/accident prevention technique. Select topics applicable to the existing work environment; plan the presentation and focus on one issue at a time. Use materials to reinforce the presentation and clarify the expected outcomes.

Why does the safety/toolbox talk have a great impact on groups? Its purposes are as follows:

- Communicate by creating understanding, educate, and add knowledge.
- Motivate by inspiring, encouraging, impelling, influencing, and developing attitudes.
- Train by changing aptitudes into abilities and improving work skills.
- Accomplish efficiency, safety, and productivity.

Thus, there are many advantages to them, such as the following:

- Save time
- Create a cooperative communication climate
- Permit individual participation
- Provide the same exposure for everyone
- Foster group spirit
- Build speaker's image

Since the safety talks have many positive effects, it is important for the speaker to do the following:

- Get attention
- Provide understanding

- Concentrate on a limited number of points
- Reiterate the key points
- Commit the audience to a future course of action
- Have a clearly defined subject
- Select those facts that will make the subject understandable and convincing to the audience
- Arrange the facts and ideas in an effective manner

No matter how effectively he/she communicates with the workforce, the employer still needs to be sure that the workforce has the competence and basic skills to perform the tasks they have been assigned. This is why employers may want to document safety talks or toolbox meetings; the company can maintain documentation to verify such efforts. It can be seen from the above narrative how effective safety/toolbox talk can be in communicating safety and health information to the workforce.

Further Readings

Reese, C.D. *Accident/Incident Prevention Techniques (Second Edition)*. Boca Raton, FL: CRC Press, 2012.

Reese, C.D. *Occupational Health and Safety Management (Third Edition)*. Boca Raton, FL: CRC Press, 2016.

Reese, C.D. and J.V. Eidson. *Handbook of OSHA Construction Safety & Health (Second Edition)*. Boca Raton, FL: CRC/Lewis Publishers, 2006.

30

Incentives/Rewards

Using an incentive as a reward can be a positive motivator, but unless behavioral change is achieved, it may be only a temporary motivator. Therefore, in order to get the behavioral change that is desired, the company needs to be aware that workers hold a more positive attitude toward their work when their supervisor provides them with the reward that they desire and is most meaningful or of value to them. Also, it has been found that employees are very receptive to rewards given to them that were not expected. Rewards of this nature seem to be even more satisfying and are received more enthusiastically than the rewards they knew they were going to receive.

Many try to use rewards or incentives to achieve the behavior changes that companies would like to see, but as can be imagined, these types of incentives/rewards have the potential to backfire. Therefore, it can be understood how very important it is that incentives and rewards be well thought out and planned so that this does not happen.

Incentives such as money are fleeting and not visually stimulating over a long period since they are quickly spent. Many companies have tried a myriad of incentives and rewards such as grocery or gas cards, gift baskets, gym memberships, mini-vacations, theater tickets, VIP parking, titles, stocks, sports tickets, free hotel lodging, weekends at a local resort, product discounts on company products or services, lotto tickets, tickets to amusement parks, catalog merchandise, or time off. If the company does not know what incentives the workforce considers most valuable, conduct a survey of the employees.

Incentives

There is no need to try to implement an incentive program if the company does not have an implemented safety and health program. All other aspects of safety and health must be in place before using an incentive program. In other words, an incentive program is a component to reinforce what presently exists. If the company does not have a structured environment in which workers sense the importance of safety and health, feel involved in the process, have safety and health-conscious leaders and supervisors as role models, and are directed by goals to prevent incidents that can cause workplace injuries and illnesses, then an incentive program is of no use.

But when an effective incentive program is an integral part of the overall workplace safety and health effort, the incentive program's goals are often to address and promote individual safety/health, peer safety/health, increase production, and reduce absenteeism. This can done by providing individuals with increasing awards for safe/healthy performance, as well as letting them select the types of products that are most valuable to them. Also, absenteeism negates the ability of an individual to receive an award. But, even if a worker had an accident during a quarter, there is still the incentive to start fresh the next quarter rather than having to wait a full year before being eligible to receive another reward; this would be demotivational. Also, as part of the program, the company hopes that peer pressure would inspire the workers to look out for each other since, as a team, they would be working for a higher reward for four quarters of safe/healthy work. The fear in using such an approach is that a fellow worker would pressure another worker to not report an injury/illness or occupationally related event. Also, in this type of program, the individual worker would be encouraged to become involved in both production and safety/health by being rewarded for giving production and safe and healthy work suggestions.

Other types of incentives or rewards that could be given are as follows:

- A title to those employees who deserve it; remember, titles cost a company nothing.
- Employee-of-the-month plaques.
- Special parking spaces.
- Personal days off (can be used at worker's discretion with approval).
- Compensatory hours.
- Money; it is good for one time and that time only.
- Bonds, stock, gold, or silver; they are usually kept and thus a constant reminder.
- Items with company logo and worker's name.
- Special commemorative items or pins.
- Tangible items (products).
- One-of-a-kind or limited-edition items (belt buckles, knives, etc.).

Incentive Programs

Incentive programs are used to reinforce workers' interest, which will stay high as long as goals are realistic and the rewards are desirable. The incentive program is result driven and is targeted to specific behaviors. It is a proactive activity or enhancement to the incentive program that stimulates achievement of goals. All employees who meet the criteria of the program are rewarded, not just select individuals receiving rewards.

Many incentive programs are going beyond goal achievement to participation. Employees are recognized for the following:

- Participating in safety activities
- Submitting safety suggestions
- Being part of safety inspections
- Reporting near misses
- Taking courses or watching safety videos
- Volunteering to be on a safety committee
- Being a member of a job safety analysis team
- Giving safety talks at safety meetings
- Fostering greater safety awareness in the workplace

Many incentive programs try to address many of the pitfalls faced by employers. This is an incentive program that tries to address individual injuries, supervisor commitment, crew peer pressure, absenteeism, and nonreporting of incidents. Use anything deemed valuable by the workforce for a reward.

One of the most powerful means of strengthening, maintaining, and changing a person's responses is the use of incentives/rewards. Other responses result in rewards and punishment, which can create barriers to human performance. When unsafe performance has been rewarded or safe performance punished, it can be expected that inappropriate performance will occur in the future. If cutting corners on safety in order to increase short-term production is rewarded, the frequency of cutting corners is going to increase. To determine if incentives have created a barrier to human performance, the question to ask is, "Has the person been rewarded for doing the job incorrectly or punished for doing it correctly?"

Summary

An example of an incentive barrier is as follows: A worker doing piecework gets a bonus for the number of parts finished over quota. The worker knows that her machine is due for scheduled maintenance, but she continues to operate it longer than recommended in order to receive more bonus money. The supervisor knows what's going on but decides not to intervene. The worker is rewarded for unsafe performance, resulting in an incentive barrier to proper performance. The worker is reinforced by the bonus money, because misuse of equipment allows more production. Note also a second thing. The supervisor ignores the unsafe behavior and, therefore, doesn't provide direction and guidance. There are times when there may be a fine line between whether the barrier is misuse of incentive or lack of direction or guidance. However, the knowledge regarding the operation of the industry should help make those determinations.

There is a tendency, when discussing incentives in industry, to think in terms of the industry's or the company's incentive/reward program. While the incentive/reward program is important, it is a small part of the concept being presented. The major concerns are those little things that occur or fail to occur after the person responds. While the opportunity for bonuses, promotions, jackets, and the like is important, events such as praise from one's supervisor or coworkers, or treatment from one's work crew, are powerful determinants of behavior. These events, which occur in normal, everyday interactions at the work site, need to be studied to see if an incentive barrier exists.

For example, an individual wears safety glasses, while the rest of the work crew does not. The crew may poke fun at the person until he/she stops wearing the glasses. Being a part of a group and getting support from the group are powerful incentives in determining performance.

The Occupational Safety and Health Administration has voiced an adverse response to using incentives or rewards since they may discriminate against workers by not giving all workers the rewards or incentives equally, thus defeating the purpose of such reward or incentive to provide motivation for those who perform their work in a safe and healthy manner.

Further Readings

Reese, C.D. *Accident/Incident Prevention Techniques (Second Edition)*. Boca Raton, FL: CRC Press, 2012.

Reese, C.D. *Occupational Health and Safety Management (Third Edition)*. Boca Raton, FL: CRC Press, 2016.

Reese, C.D. and J.V. Eidson. *Handbook of OSHA Construction Safety & Health (Second Edition)*. Boca Raton, FL: CRC/Lewis Publishers, 2006.

Section F

Hazards

This book or any written word related to occupational safety and health (OSH) deals with the release of energy, which means the existence of hazards in the workplace, for without hazards there would be no need for OSH. Addressing hazards includes their identification, understanding of the danger posed, and the analysis of the components that result in accidents or incidents. Also, this includes the inherent damage and cost of such hazardous incidents.

The contents of this section are as follows:

Chapter 31—Hazard Identification
Chapter 32—Hazard Analysis
Chapter 33—Root Cause Analysis
Chapter 34—Forms of Root Cause Analysis
Chapter 35—Cost

31

Hazard Identification

Since occupational safety and health (OSH) revolves around the presence of hazards within the work environment, it becomes a necessity to be able to identify existing hazards. Each workplace and work site has unique hazards linked to it. All employees, including management, supervisors, and workers, need to be aware of the hazards they face and take actions to mitigate their potential to cause harm.

Workers already have hazard concerns and have often devised ways to mitigate the hazards, thus preventing injuries and accidents. This type of information is invaluable when removing and reducing workplace hazards.

If workplace hazards are not identified, not only will this result in failure to understand the real risk and danger that exist, but there will be an inability to control them. This will result in accidents/incidents, injuries, illnesses, and death, which is not acceptable.

This approach to hazard identification does not require that it be conducted by someone with special training and can usually be accomplished by the use of a short/simple fill-in-the-blank questionnaire. This hazard identification technique works well where management is open and genuinely concerned about the safety and health of its workforce. The most time-consuming portion of this process is analyzing the assessment and response regarding potential hazards identified. Empowering workers to identify hazards, make recommendations on abatement of the hazard, and suggest how management can respond to these potential hazards is essential. Only three responses are required:

1. Identify the hazard.
2. Explain how the hazards could be abated.
3. Suggest what the company could do.

This is why companies and businesses need to follow an approach where

1. Inventories of some fashion are conducted that include the following:
   a. Understanding the processes
   b. Associated hazards
   c. Controls related to the hazards
   d. Department location of the process
   e. Names of all those working on each process
   f. Number of employees working on the process
   g. Medical issues that can arise from the process
   h. Historical information about the process and related hazards
2. There is a reporting system that all employees don't feel threatened when using.
3. Employees and supervisors are trained to conduct hazard identification surveys together.

Hazard identification is a process, which is controlled by management. Management must assess the outcome of the hazard identification process and determine if immediate action is necessary or if, in fact, there is an actual hazard involved. When a reported hazard is not viewed as an actual hazard, it is critical to the ongoing process to inform the worker that management does not view it as a true hazard and explain why. This will insure the continued cooperation of workers in hazard identification.

Hazard Identification for Protection

Employers must identify the workplace hazards before determining how to protect the employees. In performing work site hazard identification, consideration must include not only hazards that currently exist in the workplace but also those hazards that could occur because of changes in operations or procedures or because of other factors, such as concurrent work activities. This is why the following should occur:

1. Perform hazard identification of all work sites prior to the start of work.
2. Perform regular safety and health inspections.
3. Require supervisors and employees to inspect their workplace prior to the start of each work shift or new activity.
4. Investigate accidents and near misses.
5. Analyze trends in accident and injury data.

So that all hazards are identified, conduct comprehensive baseline work site surveys for safety and health and periodic comprehensive updated surveys. Analyze planned and new facilities, processes, materials, and equipment. Perform routine job hazard analyses.

Provide for regular site safety and health inspections, so that new or previously missed hazards and failures in hazard controls are identified. So that employee insight and experience in safety and health protection may be utilized and employee concerns may be addressed, provide a reliable system for employees, without fear of reprisal, to notify management personnel about conditions that appear hazardous and to receive timely and appropriate responses; and encourage employees to use the system. Provide for investigation of accidents and near-miss incidents, so that their causes and means for their prevention are identified. Analyze injury and illness trends over time, so that patterns with common causes can be identified and prevented.

The expected benefits of hazard identification are a decrease in the incidents of injuries, a decrease in lost workdays and absenteeism, a decrease in workers' compensation cost, better productivity, and an increase in cooperation and communication. The baseline for determining the benefit of the hazard identification can be formulated from existing company data on occupational injuries/illnesses, workers' compensation, attendance, profit, and production.

Further Readings

Reese, C.D. *Accident/Incident Prevention Techniques (Second Edition)*. Boca Raton, FL: CRC Press, 2012.

Reese, C.D. *Occupational Health and Safety Management (Third Edition)*. Boca Raton, FL: CRC Press, 2016.

Reese, C.D. and J.V. Eidson. *Handbook of OSHA Construction Safety & Health (Second Edition)*. Boca Raton, FL: CRC/Lewis Publishers, 2006.

32

Hazard Analysis

Hazard analysis is a technique used to examine the workplace for hazards with the potential to cause accidents. The information obtained by the hazard analysis process provides the foundation for making decisions upon which jobs should be altered in order for the worker to perform the work safely and expeditiously. In addition, this process allows workers to become more involved in their own destiny. For some time, involvement has been recognized as a key motivator of people. This is also a positive mechanism in fostering labor/management cooperation. This is especially true if everyone in the workplace is continuously looking for potential hazards that can result in injury, illness, or even death.

Hazard analysis is an organized effort to identify and analyze the significance of hazards associated with a process or activity. Doing a hazard analysis will hopefully identify any unacceptable risk faced when in the workplace and determine the options for managing or eliminating those risks. This is why hazard analysis can shine a light on facility or design problems, and unsafe operations that cause property damage, injuries, and even death. Also, once the problems are identified, the company can identify risk management strategies to address them.

This is why when done correctly, hazard analysis helps management and workers identify potential safety issues, discover ways to lower probability of an occurrence, and minimize the associated consequences.

With this in mind, there are many reasons to perform a hazard analysis. Some of the reasons why a hazard analysis is used are as follows:

- Be standardized and provide a uniform baseline throughout the organization
- Ensure continuity of effort and promote strong leadership
- Progressively achieve the objectives set at the beginning of the process
- Be both effective and efficient and provide a basis for capital expenditure
- Be generic enough for company-wide high- and low-level applications
- Produce fast results and have a track record of results
- Require minimal data sets and minimal data accuracy at the onset
- Be relatively inexpensive (cost effective) to implement
- Allow for both logical as well as creative thought processes
- Be flexible enough to allow for continuous improvement
- Be open ended so it is not limited
- Have no culture or demographic barriers
- Be suitable for individual as well as group application
- Lead to improved communications, motivation, and task focus
- Elevate the problem and place it into the correct context

- Be rapidly transferable to real-life issues
- Support ongoing training and research
- Identify the root cause of problem or incident

A hazard analysis is used as the first step in a process used to assess risk. The result of a hazard analysis is the identification of different types of hazards. A hazard is a potential condition and either exists or does not. It may exist on its own or in combination with other hazards (sometimes called events), and conditions become an actual functional failure, accident failure, or accident (mishap). The exact way this happens in one particular sequence is called a scenario. This scenario has a probability of occurrence. Often, a system has many potential failure scenarios. It also is assigned a classification, based on the worst-case severity of the end consequence. Risk is the combination of probability and severity. Preliminary risk levels can be provided in the hazard analysis. The validation, more precise prediction (verification), and acceptance of risk are determined by a risk assessment (analysis). The main goals of verification and analysis is to provide the best selection of means of controlling or eliminating the risk. The term risk reduction is used in several engineering specialties, including avionics, chemical process safety, safety engineering, reliability engineering, and food safety.

Hazards and Risk

A hazard is defined as a condition, event, or circumstance that could lead to or contribute to an unplanned or undesirable event. Seldom does a single hazard cause an accident or a functional failure. More often, an accident or operational failure occurs as the result of a sequence of causes. A hazard analysis will consider system state, for example, operating environment, as well as failures or malfunctions.

While in some cases, safety or reliability risk can be eliminated, in most cases, a certain degree of risk must be accepted. In order to quantify expected costs before the fact, the potential consequences and the probability of occurrence must be considered. Assessment of risk is made by combining the severity of consequence with the likelihood of occurrence in a matrix. Risks that fall into the "unacceptable" category (e.g., high severity and high probability) must be mitigated by some means to reduce the level of safety risk.

Hazard Analysis

Hazard analysis is the process of identifying hazards related to a project, process, or activities at the work site. Identify the workplace hazards before determining how to protect employees. In performing work site analyses, consider not only hazards that currently exist in the workplace but also those hazards that could occur because of changes in operations or procedures or because of other factors, such as concurrent work activities. First, perform hazard analyses of all activities or projects prior to the start of work, determine the hazards involved with each phase of the project, and perform regular safety and health site inspections.

Secondly, require supervisors and employees to inspect their workplace prior to the start of each work shift or new activity, investigate accidents and near misses, and analyze trends in accident and injury data.

When performing a hazard analysis, all hazards should be identified. This means conducting comprehensive baseline work site surveys for safety and health and periodic comprehensive updated surveys. It is important to analyze planned and new facilities, processes, materials, and equipment, as well as perform routine job hazard analyses. This also means that regular site safety and health inspections need to be conducted so that new or previously missed hazards and failures in hazard controls are identified.

A hazard reporting/response program should be developed to utilize the employees' insight and experience in safety and health protection. The employees' concerns should be addressed, and a reliable system should be provided whereby employees, without fear of reprisal, may notify management personnel about conditions that appear hazardous. These notifications should receive timely and appropriate responses, and the employees should be encouraged to use this system.

Another way to maintain a hazard analysis is to investigate accidents and near-miss incidents so that their causes and means for their prevention are identified. By analyzing injury and illness trends over a period of time, patterns with common causes can be identified and prevented.

Each company may require different types of hazard analyses, depending on the company's role, the size, the complexity of the work site, and the nature of associated hazards. Management may choose to use a project hazard analysis, a phase hazard analysis, and/or job safety assessment.

A project hazard analysis (preliminary hazard analysis) should be performed for each project prior to the start of work and should provide the basis for the project-specific safety and health plan. The project hazard analysis should identify the following:

- The anticipated phases of the project.
- The types of hazards likely to be associated with each anticipated phase.
- The control measures necessary to protect site workers from the identified hazards.
- Those phases and specific operations of the project for which activities or related protective measures must be designed, supervised, approved, or inspected by a registered professional engineer or competent person.
- Those phases and specific operations of the project that will require further analyses are those that have a complexity of the hazards or unusual activities involved; there is uncertainty concerning the site conditions that are present at the particular time to complete the phase or operation.

A phase hazard analysis may be performed for those phases of the project for which the project hazard analysis has identified the need for further analysis, and for those phases of the project for which methods or site conditions have changed since the project hazard analysis was completed. The phase hazard analysis is performed prior to the start of work on that phase of the project and is expanded, based on the results of the project hazard analysis, by providing a more thorough evaluation of related work activities and site conditions. As appropriate, the phase hazard analysis should include the following:

- Identification of the specific work operations or procedures
- An evaluation of the hazards associated with the specific chemicals, equipment, materials, and procedures used or present during the performance of that phase of work
- An evaluation of how safety and health has impacted any changes in the schedule, work procedures, or site conditions that have occurred since the performance of the project hazard analysis needs to be completed
- Identification of specific control measures necessary to protect workers from the identified hazards
- Identification of specific operations for which protective measures or procedures must be designed, supervised, approved, or inspected by a registered professional engineer or competent person

A job safety assessment or analysis should be performed at the start of any task or operation. The designated competent or authorized person should evaluate the task or operation to identify potential hazards and determine the necessary controls. When conducting any evaluation of hazards, work procedures, or processes. The unexpected should always be a possibility and a problem solving opportunity.

In addition, the authorized person shall ensure that each employee involved in the task or operation is aware of the hazards related to the task or operation and of the measures or procedures that he/she must use to protect himself/herself. Note: The job safety assessment is not intended to be a formal, documented analysis but, instead, is more of a quick check of actual site conditions and a review of planned procedures and precautions.

Hazard analysis can get sophisticated and go into much detail. Where the potential hazards are significant and the possibility for trouble is quite real, such detail may well be essential. However, for many processes and operations—both real and proposed—a solid look at the operation or plans by a variety of affected people may be sufficient.

Analysis often implies mathematics, but calculating math equations is not the major emphasis when attempting to address hazards or accidents/incidents that occur within the industry. Analysis in this context means taking time to examine systematically the work site's existing or potential hazards. This can be accomplished in a variety of ways.

However, if the company is faced with fairly sophisticated and complex risks with a reasonable probability of disaster if things go wrong, then the company may want some help with some of the other hazard analysis methodologies. What follows is a very brief look at the common ones. If the company decides to try one of the more complex approaches, check with the local Occupational Safety and Health Administration (OSHA) consultation office or call an engineering firm that specializes in hazard analysis.

Further Readings

Chiu, C. *A Comprehensive Course in Root Cause Analysis and Corrective Action for Nuclear Power Plants, Workshop Manual*. San Juan Capistrano, CA: Failure Prevention Inc., 1988.

Fine, W. "Mathematical Evaluations for Controlling Hazards," in J. Widner (Ed.). *Selected Readings in Safety*. Macon, GA: Academy Press, 1973.

Gano, D.L. "Root Cause and How to Find It," *Nuclear News*, August, 1987.

Petersen, D. *Techniques of Safety Management: A Systems Approach (Third Edition)*. Goshen, NY: Aloray Inc., 1989.

Reese, C.D. *Accident/Incident Prevention Techniques (Second Edition)*. Boca Raton, FL: CRC Press, 2012.

Reese, C.D. *Occupational Health and Safety Management (Third Edition)*. Boca Raton, FL: CRC Press, 2016.

Reese, C.D. and J.V. Eidson. *Handbook of OSHA Construction Safety & Health (Second Edition)*. Boca Raton, FL: CRC/Taylor & Francis, 2006.

United States Department of Energy, Office of Nuclear Energy. *Root Cause Analysis Guidance Document*. Washington, DC: US Department of Energy, February, 1992.

United States Department of Labor, National Mine Health and Safety Academy. *Accident Prevention Techniques*. Beckley, WV: US Department of Labor, 1984.

United States Department of Labor, Mine Safety and Health Administration. *Accident Prevention, Safety Manual No. 4*, Beckley, WV: US Department of Labor, Revised 1990.

33

Root Cause Analysis

A root cause analysis is not a search for the obvious but an in-depth look at the basic or underlying causes of occupational accidents or incidents. This is a more technical analysis and is where engineering, mathematics, and statistics begin to infiltrate occupational safety and health (OSH). This is why the following should be considered when performing analyses:

- Chart events in chronological order, developing an events and causal factors chart as initial facts become available.
- Stress aspects of the accident that may be causal factors.
- Establish accurate, complete, and substantive information that can be used to support the analysis and determine the causal factors of the accident.
- Stress aspects of the accident that may be the foundation for judgments of needs and future preventive measures.
- Resolve matters of speculation and disputed facts through investigative team discussions.
- Document methodologies used in analysis: use several techniques to explore various components of an accident.
- Qualify facts and subsequent analysis that cannot be determined with relative certainty.
- Conduct preliminary analyses: use results to guide additional collection of evidence.
- Analyze relationships of event causes.
- Clearly identify all causal factors.
- Examine management systems as potential causal factors.
- Consider the use of analytic software to assist in evidence analysis.

A root cause analysis is only the beginning and a fraction of the analysis process and should not be considered the sole approach to an analysis of an accident.

The basic reason for investigating and reporting the causes of occurrences is to enable the identification of corrective actions adequate to prevent recurrence and thereby protect the safety and health of the public, the workers, the equipment/machinery/facility, and the environment. Every root cause investigation and reporting process should include five phases. While there may be some overlap between phases, every effort should be made to keep them separate and distinct. The phases of a root cause analysis are as follows:

- Phase I—data collection
- Phase II—assessment
- Phase III—corrective actions

- Phase IV—informing
- Phase V—follow-up

The investigation process is used to gain an understanding of the occurrence, its causes, and what corrective actions are necessary to prevent recurrence. The line of reasoning in the investigation process is as follows: (1) Outline what happened step by step. (2) Begin with the occurrence and identify the problem (condition, situation, or action that was not wanted and not planned). (3) Determine what program element was supposed to have prevented this occurrence. (Was it lacking or did it fail?) (4) Investigate the reasons why this situation was permitted to exist.

Programs can then be improved and managed more efficiently and safely.

This line of reasoning will explain why the occurrence was not prevented and what corrective actions will be most effective. This reasoning should be kept in mind during the entire root cause process. Effective corrective action programs include the following:

- Management emphasis on the identification and correction of problems that can affect human and equipment performance, including assigning qualified personnel to effectively evaluate equipment and human performance problems, implementing corrective actions, and following up to verify that corrective actions are effective.
- Development of administrative procedures that describe the process, identify resources, and assign responsibility.
- Development of a working environment that requires accountability for correction of impediments to error-free task performance and reliable equipment performance.
- Development of a working environment that encourages voluntary reporting of deficiencies, errors, and omissions.
- Training programs for individuals found at fault in root cause analysis.
- Training of personnel and managers to recognize and report occurrences, including early identification of significant and generic problems.
- Development of programs to ensure prompt investigation following an occurrence or identification of declining trends in performance to determine root causes and corrective actions.
- Adoption of a classification and trending mechanism that identifies those factors that continue to cause problems with generic implications.

Summary

Determining facts related to any accident is the key to an accurate and effective analysis. This is why a root cause analysis should do the following:

- Begin defining facts early in the collection of evidence
- Develop an accident chronology (e.g., events and causal factors chart) while collecting evidence

- Set aside preconceived notions and speculation
- Allow discovery of facts to guide the investigative process
- Consider all information for relevance and possible causation
- Continually review facts to verify accuracy and relevance
- Retain all information gathered, even that which is removed from the accident chronology
- Establish a clear description of the accident

Select the one (most) direct cause and the root (basic) cause (the one for which corrective action will prevent recurrence and have the greatest, most widespread effect). In cause selection, focus on programmatic and system deficiencies, and avoid simple excuses such as blaming the employee. Note that the root (basic) cause must be an explanation (the why) of the direct cause, not a repeat of the direct cause. In addition, a cause description is not just a repeat of the category code description; it is a description specific to the occurrence. Also, up to three (contributing or indirect) causes may be selected. Describe the corrective actions selected to prevent recurrence, including the reason why they were selected, and how they will prevent recurrence. Collect additional information as necessary.

Further Readings

Chiu, C. *A Comprehensive Course in Root Cause Analysis and Corrective Action for Nuclear Power Plants, Workshop Manual.* San Juan Capistrano, CA: Failure Prevention Inc., 1988.

Gano, D.L. "Root Cause and How to Find It," *Nuclear News*, August 1987.

Nertney, R.J., J.D. Cornelison, and W.A. Trost. *Root Cause Analysis of Performance Indicators, (WP-21).* System Safety Development Center, Idaho Falls: EG&G Idaho, Inc., 1989.

Reese, C.D. *Accident/Incident Prevention Techniques (Second Edition).* Boca Raton, FL: CRC Press, 2012.

Reese, C.D. *Occupational Health and Safety Management (Third Edition).* Boca Raton, FL: CRC Press, 2016.

Reese, C.D. and J.V. Eidson. *Handbook of OSHA Construction Safety & Health (Second Edition).* Boca Raton, FL: CRC/Lewis Publishers, 2006.

United States Department of Energy. *Accident/Incident Investigation Manual, (SSDC 27, DOE/SSDC 76-45/27) (2nd Edition),* Washington, DC: US Department of Energy, November 1985.

United States Department of Energy. *Occurrence Reporting and Processing of Operations Information, (DOE Order 5000.3A).* Washington, DC: US Department of Energy, May 30, 1990.

United States Department of Energy. *User's Manual, Occurrence Reporting and Processing System (ORPS), (Draft, DOE/ID-10319).* Idaho Falls: EG&G Idaho, Inc., 1991.

United States Department of Energy, Office of Nuclear Energy. *Root Cause Analysis Guidance Document.* Washington, DC: US Department of Energy, February 1992.

34

Forms of Root Cause Analysis

Accidents are rarely simple and almost never result from a single cause. Accidents may develop from a sequence of events involving performance errors, changes in procedures, oversights, and missions. Events and conditions must be identified and examined to find the cause of the accident and a way to prevent that accident and similar accidents from occurring again. By creating an event in the causal factor chain, multiple causes can be visually illustrated, and a visual relationship between the direct and contributing causes can be shown. Event causal charting also visually delineates the interactions and relationships of all involved groups or individuals. By using root cause analysis, one can develop an event causal chain to examine the accident in a step-by-step manner by looking at the events, conditions, and causal factors chronologically, to prevent future accidents.

Root cause analysis is used when there are multiple problems with a number of causes of an accident. A root cause analysis is a sequence of events that shows, step by step, the events that took place in order for the accident to occur. Root cause analysis puts all the necessary and sufficient events and causal factors for an accident in a logical, chronological sequence. It analyzes the accident and evaluates evidence during an investigation. It is also used to help prevent similar accidents in the future and to validate the accuracy of preaccidental system analysis. It is used to help identify an accident's causal factors, which, once identified, can be fixed to eliminate future accidents of the same or of similar nature.

On the downside, root cause analysis is a time-consuming process and requires the investigator to be familiar with the process for it to be effective. Investigators may need to revisit an accident scene multiple times and look at areas that are not directly related to the accident to have a complete event and causal factor chain. Analysis requires a broad perspective of the accident to identify any hidden problems that would have caused the accident.

One of the simplest root cause analysis techniques is to determine the causes of accidents/incidents at different levels. During any hazard analysis, investigators are always trying to determine the root cause of any accident or incident. Experts who study accidents often do a breakdown or analysis of the causes. They analyze them at three different levels:

1. Direct causes (unplanned release of energy or hazardous material)
2. Indirect causes (unsafe acts and unsafe conditions)
3. Basic (root) causes (management safety policies and decisions, and personal factors)

When basic causes are eliminated, unsafe acts and unsafe conditions may not occur.

Thus, accidents have many causes. Basic (root) causes lead to unsafe acts and unsafe conditions (indirect causes). Indirect causes may result in a release of energy or hazardous material (direct causes). The direct cause may allow for contact, resulting in personal injury, property damage, or equipment failure (accident).

Root (Basic) Cause Analysis

Root causes are those that, if corrected, would eliminate the accident from occurring again or similar accidents from occurring. They may include several contributing causes. They are a higher order of causes that address multiple problems rather than focusing on the single direct cause. An example would be, "Management failed to implement the principles and core functions of a safety and health program. It is management's responsibility to ensure that the workplace has an effective safety and health program and that the workplace is safe for employees to work in."

Root Cause Analysis Methods

The most common root cause analysis methods are as follows:

1. Events and causal factor analysis identifies the time sequence of a series of tasks and actions and the surrounding conditions leading to an occurrence.

2. Change analysis is used when the problem is obscure. It is a systematic process that is generally used for a single occurrence and focuses on elements that have changed.

3. Barrier analysis is a systematic process that can be used to identify physical, administrative, and procedural barriers or controls that should have prevented the occurrence.

4. Management oversight and risk tree (MORT) analysis is used to identify inadequacies in barriers and controls, specific barrier and support functions, and management functions. It identifies specific factors relating to an occurrence and identifies the management factors that permitted these risk factors to exist.

5. Human performance evaluation identifies factors that influence task performance. The focus of this analysis method is on operability, work environment, and management factors. User–system interface studies are frequently done to improve performance. This takes precedence over disciplinary measures. Human performance evaluation is used to identify factors that influence task performance.

6. Kepner-Tregoe (K-T) problem solving and decision making provides a systematic framework for gathering, organizing, and evaluating information and applies to all phases of the occurrence investigation process. Data are needed for those using this method. A further description is not included in this book.

The use of different methods to conduct root cause analysis has been widely accepted over a period of years. There have been many creative adaptations and permutations using the root cause analysis approach, but the foundation for it has stood the test of time. Certain methods are used for different circumstances, such as when they fit well for certain industries, for unique hazards, when engineering becomes a factor, or when complexity is present. The analysis of an accident does not stop with the identification of the direct, indirect, and basic (root) causes of the accident or incident. To make positive gains from the event, changes should be made in the interaction of users, systems, materials, methods, and physical and social environments. These changes should result from the recommendations that are derived from the causes identified during the investigation. The goal of these changes is the prevention of future accidents and incidents similar to the one investigated.

This type of analysis may trigger the need to more closely analyze a job or task that has been identified as presenting a high risk of producing hazards or injuries.

System Safety Engineering

System safety is the use of a combination of management and systems engineering techniques that is integrated into the evaluation and reduction of risk in a system, operation, or process. The overall purpose is to identify hazards, eliminate or control them, and mitigate any remaining risk. It should be a comprehensive approach to managing risk. Many techniques have been developed for this purpose. A number of these are listed in the following:

- **Time loss analysis:** Time loss analysis evaluates emergency response performance.
- **Human factor analysis:** Human factor analysis identifies elements that influence task performance, focusing on operability, work environment, and management elements.
- **Integrated accident event matrix:** An integrated accident event matrix illustrates the time-based interaction between the victim and other key personnel prior to the accident and between the emergency responders and the victim after the accident.
- **Failure modes and effects analysis:** This method is most often used in the hazard analysis of systems and subsystems; it is primarily concerned with evaluating single-point failures, probability of accidents or occurrences, and reliability of systems and subsystems.
- **Software hazard analysis:** This analytic technique is used to locate software-based failures that could have contributed to an accident.
- **Common cause failure analysis:** Common cause failure analysis evaluates multiple failures that may be caused by a single event shared by multiple components.
- **Sneak circuit analysis:** A sneak circuit is an unanticipated energy path that can enable a failure, prevent a wanted function, or produce a mistiming of system functions. Sneak circuit analysis is mainly performed on electronic circuitry, but it can also be used in situations involving hydraulic, pneumatic, mechanical, and software systems.
- **Materials and structural analysis:** Materials and structural analysis is used to test and analyze physical evidence.
- **Design criteria analysis:** This method involves the systematic review of standards, codes, design specifications, procedures, and policies relevant to the accident.
- **Accident reconstruction:** Although not widely used in accident investigations, accident reconstruction may be useful when accident scenes yield sketchy, inconclusive evidence.

- **Scientific modeling:** Scientific modeling models the behavior of a physical process or phenomenon. The methods, which range from simple hand calculations to complex and highly specialized computer models, cover a wide spectrum of physical processes (e.g., nuclear criticality, atmospheric dispersion, groundwater and surface water transport/dispersion, nuclear reactor physics, fire modeling, chemical reaction modeling, explosive modeling).

These more sophisticated analyses become the realm of science and engineering or those having specific training or expertise and should not be undertaken lightly or as a cure to all existing problems. This is an occasion when complexity becomes more than simple occupational safety and health (OSH) and usually requires the use of outside expertise.

Further Readings

Chiu, C. *A Comprehensive Course in Root Cause Analysis and Corrective Action for Nuclear Power Plants, Workshop Manual.* San Juan Capistrano, CA: Failure Prevention Inc., 1988.

Gano, D.L. "Root Cause and How to Find It," *Nuclear News*, August 1987.

Nertney, R.J., J.D. Cornelison, and W.A. Trost. *Root Cause Analysis of Performance Indicators, (WP-21).* System Safety Development Center, Idaho Falls, ID: EG&G Idaho, Inc, 1989.

Reese, C.D. *Accident/Incident Prevention Techniques (Second Edition).* Boca Raton, FL: CRC Press, 2012.

Reese, C.D. *Occupational Health and Safety Management (Third Edition).* Boca Raton, FL: CRC Press, 2016.

Reese, C.D. and J.V. Eidson. *Handbook of OSHA Construction Safety & Health (Second Edition).* Boca Raton, FL: CRC/Lewis Publishers 2006.

United States Department of Energy. *Accident/Incident Investigation Manual, (SSDC 27, DOE/SSDC 76-45/27) (2nd Edition).* Washington, DC: Department of Energy, November 1985.

United States Department of Energy. *Occurrence Reporting and Processing of Operations Information,* (DOE Order 5000.3A). Washington, DC: Department of Energy, May 30, 1990.

United States Department of Energy. *User's Manual, Occurrence Reporting and Processing System (ORPS), (Draft, DOE/ID-10319).* Idaho Falls, ID: EG&G Idaho, Inc, 1991.

United States Department of Energy, Office of Nuclear Energy. Root Cause Analysis Guidance Document. Washington, DC: Department of Energy, February 1992.

35

Cost

Cost is always the driving force for companies. Anything that involves occupational safety and health (OSH) has cost linked to it. This is why companies want to know the following:

- What is safety and health costing?
- How much does an accident cost?
- Is the company avoiding cost?
- Is compliance saving money?
- Is prevention saving money?
- Are our safety and health dollars cost effective?
- What are the benefits of safety and health related to dollars?
- Does safety and health avoid other cost?

The True Bottom Line

Without exception, all industries and companies face safety and health issues, which could have adverse effects upon their workforce and workplace. It matters not whether the company is a service industry, insurance agency, construction operation, or manufacturer of widgets. The workforce will be exposed to the hazards unique to that work site. It is definitely beneficial to the bottom line to not have any of the workforce injured or ill from something within that workplace. Whether the businesses are large or small, having anyone in their workforce who has been incapacitated in any way disrupts the work process. Not counting the loss of a potential key employee, the time spent addressing an incident that has caused injuries or illnesses definitely cuts into the bottom line. If this seems bad, an occupation death really has an impact.

During the investigation of an occupational death, companies often cannot function and fold because they cannot absorb the impact and cost of a job-related death. The cost alone for such an occurrence exceeds $1,200,000. If a small company experiences such an event, a workplace fatality can be business-ending.

Taking a reasonable amount of time to address OSH will have a very positive impact upon a business's or company's particular operation. Certainly, the magnitude of the safety and health effort will vary depending upon whether the workplace is an office environment or a construction jobsite. If a company just addresses the key components of OSH, they will have made some great strides forward in making safety and health an integral part of the workplace and a concrete demonstration as to the value gained by investing real dollars in occupational safety and health.

Cost of Accidents

The direct (insured) cost of accidents is by far the easiest to track. These direct costs are for medical care, repairing or replacing damaged equipment, and workers' compensation premiums. There is no possible way to not see the cost of an ambulance, hospital bill, or repair bill when it comes. Likewise, employers know the dollar amounts being expended on workers' compensation.

Data from the National Safety Council for 2008 indicate that the cost of work-related injuries and deaths was $183.0 billion. Wage and productivity loss accounted for $88.4 billion, medical cost $38.3 billion, and employer cost equaled $12.7 billion. The average cost of a workplace death was put at $1,310,000, and a disabling injury cost $48,000. A look at other injury costs provided by the National Safety Council indicates that a reasonable, serious, nondisabling injury would have an average cost of $22,674 (2006–2007).

Many safety and health experts estimate that the indirect (uninsured) cost of accidents, and the costs associated with them, equal 5 to 10 times the direct cost of the accidents. These indirect costs are caused by many of the following:

- Time lost from work by the injured.
- Loss in earning power.
- Economic loss to injured worker's family.
- Lost time by fellow workers.
- Loss of efficiency due to breakup of crew.
- Lost time of supervisor.
- Cost of breaking in a new worker.
- Damage to tools and equipment.
- The time damaged equipment is out of service.
- Spoiled work.
- Loss of production.
- Spoilage from fire, water, chemical, explosives, etc.
- Failure to fill orders.
- Overhead cost (while work was disrupted).
- Miscellaneous—there are at least 100 other items of cost that appear one or more times with every accident.

Summary

Obtaining the cost of accidents continues to be an issue facing safety and health personnel, management, policy makers, and researchers. Estimated costs are used to set priorities, make program decisions, make business decisions, make personnel decisions, support budget decisions, as well as make other cost-affecting decisions. These decisions and other activities are based upon the accuracy of the data on the company's safety and health record. Accident/incident records and the cost of providing for the safety and health of employees must be accurate and reliable. A good safety and health accounting system serves as the basis for making informed and reliable management decisions.

Further Readings

Reese, C.D. *Accident/Incident Prevention Techniques (Second Edition)*. Boca Raton, FL: CRC Press, 2012.

Reese, C.D. *Occupational Health and Safety Management (Third Edition)*. Boca Raton, FL: CRC Press, 2016.

Reese, C.D. and J.V. Eidson. *Handbook of OSHA Construction Safety & Health (Second Edition)*. Boca Raton, FL: CRC/Lewis Publishers, 2006.

United States Department of Labor, Occupational Safety and Health Administration, Office of Training and Education. *OSHA Voluntary Compliance Outreach Program: Instructors Reference Manual*. Des Plaines, IL: US Department of Labor, 1993.

Section G

Risk

When hazards exist in the workplace, there is the danger that workers could be exposed to them. The degree of risk to the worker needs to be determined. The question of how much risk is deemed acceptable and how much actual risk the worker is confronted with needs to be assessed. Also, since zero risk is not always possible, then risk must be managed in order to control the workplace risk potential.

The contents of this section are as follows:

Chapter 36—Acceptable or Tolerable Risk

Chapter 37—Risk Assessment

Chapter 38—Risk Management

36

Acceptable or Tolerable Risk

As is known, life can never be risk-free. Since zero risk is unachievable, a risk assessment is necessary to mitigate or reduce the risk, which effectively reduces the danger imposed by the risk. It suffices to state that all risk carries with it potential danger from injury, illness, and death in the workplace and, under most circumstances, cannot be removed or eliminated; neither can the risk be reengineered out or can a substitute for it be found in all cases. Workplaces have hazards that present a risk of injury or illness from the dangers that exist. At times, the hazards cannot be removed, and the dangers exist and can result in an accident. Risk is the probability of an accident occurring. The amount of risk that is deemed acceptable or that can be tolerated will do much to define the extent of the injury prevention effort.

A determination needs to be made in assessing the amount of acceptable risk in the workplace from hazards, which present a risk of injury or illness from the dangers that exist. At times, the hazards cannot be removed, and dangers exist that could result in an accident, event, or incident. Risk related to safety and health is often a judgment call. But, even a judgment call can be quantified if criteria are developed and a value is placed upon them.

Why Can Risk Be Tolerated at an Acceptable Level?

The following is a list of why acceptable risk can exist:

- It falls below an arbitrary defined probability.
- It falls below some level that is already tolerated.
- It falls below the arbitrary defined attributable fraction of total accidents or illnesses in the workplace.
- The cost of reducing the risk would exceed the cost saved.
- The cost of reducing risk would exceed the costs saved when "cost of suffering" is also factored in.
- The opportunity cost would be better spent on other more pressing safety and health issues.
- Professional occupational safety and health professionals say it is acceptable.
- The workers deem the risk acceptable.
- The people in the community deem it acceptable.
- Politicians say it is acceptable.
- Company management deems it acceptable.

In the public domain, a risk of one in a million is considered acceptable risk (e.g., cancer, radiation poisoning, trauma death, automobile injuries, etc.), while in the workplace, a fence line mentality exists that accepts 1 in 1,000 occupation injuries, illnesses, or deaths. Risk acceptance always seems to be higher for workers. It is assumed that workers know and accept the risk of working, as do their employers. Employers are bottom line oriented and assume that acceptable or tolerated risk is alright if the economic saving arising out of action to reduce a risk outweighs the cost of such an action. In business, cost–benefit is a type of analysis that considers the cost of physical damage, cost to repair, cost to replace, cost to business's or company's reputation, liability/legal cost, and cost in relation to the human toll.

However, even judgment calls can be quantified if you develop criteria and place value upon them. W. Fine has provided a mathematical model (which has been updated by others from time to time) for conducting a risk assessment that results in a numerical value, which can be used to compare potential risks from accidents as well as determine if the amount of fixing justifies the cost involved to fix or remove the hazard. Most of his basic components are present, which will allow for assessing the risk of a hazard as well as making decisions on whether it is logical and economically feasible to fix the hazard.

In determining risk, a consequence value needs to be developed for the existing hazard, which indicates its effect on workers if they come into contact with it, as well as a value for the exposure potential, which denotes how many times during any period workers could come into contact with the hazard. There is also a need to assess the probability that workers would be injured, become ill, or be killed if they came into contact with the hazard. We use these three criteria since they are variables:

- Consequences—ranging from minor first-aid injuries to serious (amputation) or very serious (disabling) injuries, injury to multiple workers, or death.
- Exposure—ranging from minutes to hours, days, weeks, months, or years.
- Probability—ranging from 0% to 100%.

Factored together, these three allow for a determination of the amount of risk posed by a hazard. This is not a determination of acceptable risk. No matter the business or operation, another question that arises from this score is, "Should we invest the money to fix or remove the hazard, and what will be gained?" Another way to represent this is, "How much fixing will be gotten for the money invested?" Companies are always looking at the bottom line. The two factors in determining the justification factor for fixing or removing a hazard are dependent on the cost and the amount of fixing or removal of the hazard. This justification factor is obtained by determining the amount of fixing or removal (ranging from 100% to 0%). Secondly, what is the cost of fixing or removal (ranging from <$100 to >$1,000,000)? In making a business decision, one should consider whether the factor justifies the fix.

This is only a process that can be amended, not used, or used as one tool to help prioritize the tolerable or acceptable risk the company is willing to endure. It will allow for generating numbers to compare potential hazards and make informed decisions on addressing the existing hazard. When trying to make a case to management to fund safety/health-related items, quantitative approaches are better received than ones based on opinions or worded facts. Statistical data always help to bolster the case for doing almost anything.

In summary, determining acceptable risk is a complicated decision-making process, which means addressing more than safety and health only but also encompassing the areas of organization, management, employees, social issues, and political issues for input to the final decision-making process regarding how much occupational safety and health (OSH), how much risk acceptance, and the tolerance for risk.

Further Readings

Fine, W. "Mathematical Evaluations for Controlling Hazards," in J. Widner (Ed.). *Selected Readings in Safety*. Macon, GA: Academy Press, 1973.

Petersen, D. *Techniques of Safety Management: A Systems Approach (3rd Edition)*. Goshen, NY: Aloray Inc., 1989.

Reese, C.D. *Accident/Incident Prevention Techniques (Second Edition)*. Boca Raton, FL: CRC Press, 2012.

Reese, C.D. *Occupational Health and Safety Management (Third Edition)*. Boca Raton, FL: CRC Press, 2016.

Reese, C.D. and J.V. Eidson. *Handbook of OSHA Construction Safety & Health (Second Edition)*. Boca Raton, FL: CRC/Lewis Publishers, 2006.

37

Risk Assessment

Risk assessment is the overall effect of the potential outcome of the release of hazardous energy in a workplace. These risks can include health risk from chemical exposure, environmental factors culminating in illnesses/disease, or physical factors that can result in traumatic injuries or death.

Risk assessment, although generic by name, becomes very specific when specific risks are addressed, such as security, a specific process, an explosive situation, a toxic chemical, or an energy-producing incident.

Developing a Risk Assessment

Thus, risk assessment is often subjective even when it is completed using as much quantitative data as is available. With this in mind, it should address four key components that may or may not be quantifiable.

The steps that need to be undertaken are as follows:

First, a walkthrough of the workplace. This may result in a few identifiable hazards or a hundred. This of course varies greatly from workplace to workplace. This hazard identification and hazard ranking is the first step in a risk assessment.

Secondly this would be followed by determining the exposure level versus the actual effects of that exposure level. This allows for a determination of what the outcomes of such an exposure would be for the employee or groups of employees.

Thirdly, these degrees of potential effects guide the determination as to the real danger or perceived risk. Judgment as to risk is evaluated based on amount of risk from potential exposure, results of exposure, how likely the exposure is to end in an illness or injury, and what the magnitude of an event is from the release of errant/uncontrolled source of energy.

The previous come together to calculate a risk assessment factor that may range along the continuum from extreme risk to low (negligible) risk.

Fourth, a risk assessment is confounded by the business side of the equation, such as how much it will cost to fix or remove a risk. The other question that needs to be answered is whether this is a justifiable business decision. Is it worth the money for the amount of fixing obtained or the cost to remove the risk?

This is why the quality of the risk assessment is vital to making both human and business decisions from a position of knowledge and information.

Purpose of Risk Assessment

The purpose of risk assessment is as follows:

- Identify existing hazards.
- Determine the extent of danger that exists.
- Provide for the ranking of hazards.
- Organize information for decision making.
- Give guidance for decision making.
- Allow for prioritizing risks.
- Help in characterizing risks, the potential safety and health effects of workers exposed to chemical, physical, and environmental hazards.
- Identify whether a safety or health hazard dose–response is a critical component to be included.
- The risk assessment is a needed element of the risk management process.

Even when a risk assessment is conducted, it may result in, at best, a guesstimate due to the lack of quantitative data. This is especially true when trying to establish a real risk picture of a carcinogen. This is because a safe level of exposure cannot be determined.

Risk assessments have become an important component in assuring a safe and effective operation. As safety is designed into processes, risk assessment becomes increasingly important. This will allow engineers to make good designs and operational decisions.

Explaining the Risk Assessment

When trying to convey the outcome of a risk assessment, it must be clearly presented, represent reality to both individuals and work groups, and be based upon as much hard data as possible. Secondly, it must be perceived as inherently useful for decision making. Thirdly, it must be credible and defensible in order for it to be acceptable, useable, and applicable.

A risk assessment must be logical, but it still may not be acceptable if based upon a foregone assumption of the risk. The risk assessment is not finite since new data and changes in processes occur, or there is a need to further reduce risk.

Risk Evaluation

Prior to any risk evaluation, a risk assessment must exist first. A risk evaluation is taking the information available and using it to make decisions. It also incorporates probability. It is a risk-ranking process using probability as its mainstay.

Summary

Risk assessment is utilized to determine the amount or degree of potential danger perceived by a given worker when deciding if an appropriate amount of action has taken place to reduce the risk for that individual to accomplish a task in a safe and healthy manner.

A risk assessment utilizes information, a hazard analysis method, and an intervention/control program evaluation that allows for an estimation of the amount of risk.

Further Readings

Bahr, N.J. *System Safety Engineering and Risk Assessment: A Practical Approach.* New York: Taylor & Francis, 1997.

Burns, T.E. *Serious Incident Prevention (Second Edition).* Boston, MA: Gulf Professional Publishing, 1946.

Daugherty, J.E. *Industrial Safety Management: A Practical Approach.* Rockville, MD: Government Institutes, 1999.

Goetsch, D.L. *Occupational Safety and Health for Technologists, Engineers, and Managers (Fifth Edition).* Upper Saddle River, NJ: Pearson Prentice Hall, 2005.

Kohn, J.P. and T.S. Ferry. *Safety and Health Management Planning.* Rockville, MD: Government Institutes, 1999.

Lack, R.W. *Safety, Health, and Asset Protection: Management Essentials (Second Edition).* Boca Raton, FL: Lewis Publishers, 2002.

38

Risk Management

The management of risk is similar to all other attempts to manage processes, systems, and other safety and health issues. As with anything requiring managing, some elements are somewhat universal when devising a management approach. Many of the critical parts (elements) have been discussed in previous chapters of this book. Thus, the requirements for managing risk include many time-tested and successful endeavors. The saying that there is nothing new under the sun is true regarding risk management. The elements of such a program usually require tweaking of the foundational components. The components suggested in these pages can be revised, added to, or removed and still provide a workable effort. The suggestions within these pages are no more than a template for developing, improving, and implementing a risk management program.

Risk Management Components

All occupational safety and health (OSH) programs begin with the critical component of management commitment and leadership. There is no use to even proceed from this point without the aforementioned element existing since all resources, support, and authority come from this element of the effort.

Secondly, management's dictate to employees is not a solution to the goals and intent of management's undertaking no matter the good intention of any attempt to improve the work environment. Failure to include employees in developing, implementing, maintaining, and improving the overall goal of the program decreases its relevance to the workforce. Buy-in by employees is critical to success. Management needs the experience, knowledge base, and ownership in order to proceed toward leveraging the power employees possess to make management's intent successful.

Thirdly, there must be an understanding of the risk that a company faces even if no incidents or events have transpired. It is important that a risk evaluation has been completed using a risk assessment, which will help rate and rank potential risk events. Before risk can be managed, it must be recognized and accepted as having the potential to cause extensive damage (catastrophic), financial impact, long-term implications, and real business impacts, all of which can possibly threaten a business's existence.

Fourth, with risk being understood and accepted as exhibiting a potential hazard that could result in an incident with damage and loss of life, businesses must take steps to control and prevent these types of events. It is easily realized the some types of risk, due to their outcomes, require more attention than others. Some efforts to control risk are rather simplistic, while others will require larger allocation of resources. The failure to implement appropriate controls suggests the employer either does not care or fails to understand the risk involved in his/her workplace. Failure to control risk is the path to, ultimately, a businesses' failure.

Fifth, the failure of operators' performance can be considered a risk to any company process. The failure to perform may result from poor operator skill, failure of training, or poor workplace design. The development of precise performance standards is critical to a safe/healthy and productive workplace. Assuring that performance is high quality and high expectations are part of the management of risk.

Without having explicit/detailed performance standards, managers, supervisors, and employees make their own decisions on what is right versus wrong, safe versus unsafe, and good versus bad without criteria based on research, data, written directions, rules of operation, and acceptable practices, and instead rely on personal decision making. The types of performance standards that are most effective dwell on both incident prevention and resource utilization.

Performance standards should provide a clear explanation of expectations of operation, performance, and outcomes that must be managed.

Sixth, a management system is one of checks and balances, and risk must be managed similarly to anything else managed by a company. It is critical that there be a measurement system that tracks progress and identifies shortcomings, needs for improvement, and achievements. This information will provide tangible data that can be utilized to provide feedback that can facilitate the improvement and upgrading of the risk management process. If feedback is a constructive process, aimed at system improvement and upgrade, rather than a punishment or personal degradation approach, then it will be accepted as an integral part of the management approach.

Seventh, it is imperative that in this process, the positive be reinforced to strengthen what is going right and steps be taken to devise and implement corrective action. But, this is not a stopping step. It is the next step that culminates this endeavor.

Eighth, in the management trend of the day, the goal is continuous improvement (lean safety). This entails the use of all existing resources to constantly strive to improve and update all processes. This includes risk management as a part of accomplishing this goal.

Why Manage Risk?

It is important to manage risk since it provides so may important and positive impacts upon occupational safety and health in the workplace. Some of its contributions to risk management are as follows:

- Management is a process, not an idea.
- It helps control or decrease risk.
- It demonstrates the value placed on safety and health.
- It organizes a process by providing structure.
- It gives concrete specific directions and guidance to the process.
- It provides standards under which everyone functions.
- It is an inclusive process.
- It fosters involvement and participation.
- It is a motivational tool.

Summary

A failure to manage an important component only suggests that the intent is to manage to fail. Managing risk is too critical to safety and health in the workplace to not do it.

Further Readings

Bahr, N.J. *System Safety Engineering and Risk Assessment: A Practical Approach*. New York: Taylor & Francis, 1997.

Burns, T.E. *Serious Incident Prevention (Second Edition)*. Boston, MA: Gulf Professional Publishing, 1946.

Daugherty, J.E. *Industrial Safety Management: A Practical Approach*. Rockville, MD: Government Institutes, 1999.

Goetsch, D.L. *Occupational Safety and Health for Technologists, Engineers, and Managers (Fifth Edition)*. Upper Saddle River, NJ: Pearson Prentice Hall, 2005.

Kohn, J.P. and T.S. Ferry. *Safety and Health Management Planning*. Rockville, MD: Government Institutes, 1999.

Lack, R.W. *Safety, Health, and Asset Protection: Management Essentials (Second Edition)*. Boca Raton, FL: Lewis Publishers, 2002.

Section H

Controlling Hazards

In addressing hazards, it is noted that most occupational safety and health (OSH) accident/incidents often result in injuries, illnesses, or deaths that are caused by a failure to control the energy released when a control fails or fails to exist. Coming in contact with energy in motion or stored energy results in an event that could have been prevented if an appropriate or effective control were present.

The contents of this section are as follows:

Chapter 39—Designing for Prevention

Chapter 40—Controls

Chapter 41—Personal Protective Equipment

Chapter 42—Occupational Safety and Health Administration (OSHA)

39

Designing for Prevention

As a way of controlling hazards, the term *designing out* can come into the picture. What this means is that instead of controlling hazards, why not prevent them from occurring in the first place? This is visualized as designing for safety and health by designing out any potential hazards before they become they become part of a process or become a machine or piece of equipment that has an inherent risk or hazard.

This would be like designing a workplace in the same fashion as designing a spaceship. From experience, it is known that designing out safety and health risk starts before the actual paper-and-pencil design stage. It begins with the idea and with the stated goal of having no safety or health issues to be found in the final design of the product or process. At times, this means the need for redundant systems, purchasing requirements, fail-safe devices, good quality of material used, and expressed warranties.

This means that human factors relevant to operating equipment are a part of the design such that controls are appropriate for the operator whether operator is left or right handed. The equipment or machines' process in the direction operators expressly view as normal. Any change in normal operating procedures increases the risk of an unplanned event.

Designing for Humans

The mission is the prevention of hazards while removing the exposure. Primarily, the intent is to protect workers. This can be accomplished in a number of ways:

- Remove the worker by using an automated system or robots.
- Remove the employee by placing him/her in a remote/enclosed operator control room.
- Design out the hazard.
- Design an operator area that fits the operator both physically and environmentally.
- Make operation a trainable function.
- Design for human factors (make operation fit the majority). Do not place a worker in an untenable position, such as placing a one-armed worker in a two-arm-requirement operation.
- If safety and health cannot be designed into an operation, find another way of doing the operation. This is a real commitment to designing for prevention.

Designers' Responsibility

These are the steps just prior to engineering design. The engineers are responsible for the design of processes, systems, machines, and equipment. As a part of the design process, they are charged with designing and incorporating safety and health into their designs.

If they cannot remove the risk in the design phase, then the effort must be to design the least exposure to workers as possible. At times, this can be accomplished by using enclosures or perimeter fencing. If a worker enters points of operations, there need to be interlocks or automatic-shut devices such as presence-sensing devices designed into the system to disengage any operation. Also, warning devices can be incorporated to draw workers' attention to potential risk from hazards.

Designing out hazards is a responsible part of doing business. This will necessitate that a company, business, or organization require, as a condition of purchasing, safety and health to be an integral part of the purchasing agreement. This is before the design process, to foster better control of potential hazards. There must be a dedication to reducing risk as a way of doing business in a safe and healthy manner.

Further Readings

MacLeod, D. *The Ergonomics Edge: Improving Safety, Quality, and Productivity.* New York: Van Nostrand Reinhold, 1995.
McCormick, E.J. *Human Factors in Engineering and Design.* New York: McGraw-Hill, 1976.

40

Controls

Ideally, hazards should be controlled by applying modern management principles. Use a comprehensive, proactive system to control hazards rather than a reactive, piecemeal response to each concern as it arises. This chapter discusses why an employer should take a proactive approach to do the following:

- Evaluate the hazard needing control.
- Eliminate the hazard.
- Select the best control available.
- Use a temporary control until a more effective one can be implemented.
- Check the effectiveness of the control.
- Reassess as necessary.

Technical Aspect of Hazard Controls

As a first step in hazard control, determine if the hazards can be controlled at their source (where the problem is created) through applied engineering. If this does not work, try to put controls between the source and the worker. The closer a control is to the source of the hazard, the better. If this is not possible, hazards must be controlled at the level of the worker. For example, workers can be required to use a specific work procedure to prevent harm.

One type of hazard control may not be completely effective. A combination of several different types of hazard controls often works well. Whatever method is used, an attempt should be made to try to find the root cause of each hazard and not simply control the symptoms. For example, it might be better to redesign a work process than simply improve a work procedure. It is better to replace, redesign, isolate, or quiet a noisy machine than to issue nearby workers hearing protectors. There are many potential mechanisms for controlling hazards, such as the following:

Source Control

- Elimination
- Substitution
- Redesign
- Isolation
- Ventilation
- Lockout
- Automation

Control along the Path from the Hazard to the Worker

Hazards that cannot be isolated, replaced, enclosed, or automated can sometimes be removed, blocked, absorbed, or diluted before they reach workers. Usually, the further away a control keeps hazards from workers, the more effective it is.

Barriers

A hazard can be blocked. For example, proper equipment guarding can protect workers from coming into contact with moving parts. Screens and barriers can block welding flash from reaching workers. Machinery lockout systems can protect maintenance workers from physical agents such as electricity, heat, pressure, and radiation.

Absorption

Baffles can block or absorb noise. Local exhaust ventilation can remove toxic gases, dusts, and fumes where they are produced.

Dilution

Some hazards can be diluted or dissipated, for example, general (dilution) ventilation.

Control at the Level of the Worker

Control at the level of the worker usually does not remove the risk posed by a hazard.

Selecting Controls

Selecting a control often involves evaluating and selecting temporary and permanent controls, implementing temporary measures until permanent (engineering) controls can be put in place, and implementing permanent controls when reasonably practicable. For example, suppose a noise hazard is identified. Temporary measures might require workers to use hearing protection. Long-term, permanent controls might use engineering to remove or isolate the noise source.

Risk Control

The key to risk control is to prevent exposure to those who could be at risk. In the workplace, it is not possible to have no exposure if anything is going to get done. It is important to limit the potential exposure or amount of exposure. The basic principles of protection from radiation exposure provide the foundation for risk control. The three elements of exposure control are distance, time, and shielding. Distance provides the best mechanism to prevent exposure. Distance can be physical distance or remote distance where robotics can provide the distance and limit the exposure. Time is an exposure limiter and a mechanism that allows employers to spread exposure over several workers. The time period of the work cycle that allows for minimal risk exposure to transpire is during a shift when fewer employees are present to be exposed. Often, second and third shifts are times when fewer workers are present. Shielding is frequently considered the least acceptable approach to risk control. Barriers or PPE should be the risk control of last resort. With occupational safety and health (OSH), other approaches to risk control are often employed.

Evaluating the Effectiveness of Controls

Sometimes, hazard controls do not work as well as expected. Therefore, the committee or representative should monitor the effectiveness of the corrective action taken by the employer during inspections and other activities. Ask these questions:

- Have the controls solved the problem?
- Is the risk posed by the original hazard contained?
- Have any new hazards been created?
- Are new hazards appropriately controlled?
- Are monitoring processes adequate?

- Have workers been adequately informed about the situation?
- Have orientation and training programs been modified to deal with the new situation?
- Are any other measures required?
- Are the effectiveness of hazard control documented in your committee minutes?

Awareness Devices

Awareness devices are linked to the senses. They are warning devices that can be heard and seen. They act as alerts to workers but create no type of physical barrier. They are found in most workplaces and carry with them a moderate degree of effectiveness. The following are such devices:

- Backup alarms
- Warning signals, both audible and visual
- Warning odor
- Beepers
- Horns
- Sirens
- Labels
- Warning signs

Work Practices

Work practices concern the ways in which a job task or activity is done. This may mean that you create a specific procedure for completing the task or job. It may also mean that you implement special training for a job or task.

Administrative Controls

These include introducing new policies, improving work procedures, and requiring workers to use specific PEE and hygiene practices. For example, job rotations and scheduling can reduce the time that workers are exposed to a hazard. Workers can be rotated through jobs requiring repetitive tendon and muscle movements to prevent cumulative trauma injuries. Noisy processes can be scheduled when few workers are in the workplace. Standardized written work procedures can ensure that work is done safely. Employees can be required to use shower and change facilities to prevent absorption of chemical contaminants. The employer is responsible for enforcing administrative controls.

A second approach is to control the hazard through administrative directives. This may be accomplished by rotating workers, which allows you to limit their exposure, or having workers only work in areas when no hazards exist during that part of their shift. This applies particularly to chemical exposures and repetitive activities that could result in ergonomic-related incidents. Examples of administrative controls are as follows:

- Requiring specific training and education
- Scheduling off-shift work
- Worker rotation

Management controls are needed to express the company's view of hazards and their response to hazards that have been detected. The entire program must be directed and supported through the management controls. If management does not have a systematic and set procedure for addressing the control of hazards in place, the reporting/identifying of hazards is a waste of time and dollars. This goes back to the policies and directives and the holding accountable of those responsible by providing them with the resources (budget) for correcting and controlling hazards. Some aspects of management controls are as follows:

- Policies
- Directives
- Responsibilities (line and staff)
- Vigor and example
- Accountability
- Budget

The attempt to identify the work site hazards and address them should be an integral part of your management approach. If the hazards are not addressed in a timely fashion, they will not be identified or reported. If dollars become the main reason for not fixing or controlling hazards, you will lose the motivation of the workforce to identify or report them.

Work Procedures, Training, and Supervision

Supervisors can be trained to apply modern safety management and supervisory practices. Workers can be trained to use standardized safe work practices. The employer should periodically review and update operating procedures and worker training. Refresher training should be offered periodically. The employer is expected to ensure that employees follow safe work practices.

Hazard Prevention and Controls

The Occupational Safety and Health Administration (OSHA) requires employers to protect their employees from workplace hazards such as machines, work procedures, and hazardous substances that can cause injury or illnesses. It is known from past practices and situations that something must be done to mitigate or remove hazards from the workplace. Actions taken often create other hazards that did not exist before attempting to address the existing hazard.

Many companies have suggestion programs where workers receive rewards for suggestions that are implemented. It is no secret that the person who often has the best ideas on how to decrease or remove a hazard is the one who faces that hazard as part of doing normal work. Involving those who are impacted most in decision-making processes that affect their work is a sound management practice.

Hazard Control Summary

All identified hazardous conditions should be eliminated or controlled immediately. Where this is not possible, interim control measures are to be implemented immediately to protect workers, warning signs must be posted at the location of the hazard, all affected employees need to be informed of the location of the hazard and of the required interim controls, and permanent control measures must be implemented as soon as possible.

Controls come in all forms, from engineering devices and administrative policy to PPE. The best controls can be placed upon equipment before involving people and, thus, either preclude or guard the workforce from hazards. Administrative controls rely upon individuals following policies, guidelines, and procedures to control hazards and exposure to hazards. However, as we all know, this certainly provides no guarantee that the protective policies and procedures will be adhered to unless effective supervision and enforcement exist. Again, this relies on the company having a strong commitment to OSH

The use of PPE will not control hazards unless individuals who are exposed to the hazards are wearing the appropriate PPE. The use of PPE is usually considered the control of last resort since it has always been difficult for companies to be sure that exposed individuals are indeed wearing the required PPE.

Where a supervisor or foreman is not sure how to correct an identified hazard or is not sure if a specific condition presents a hazard, he/she shall seek technical assistance from the designated competent person, safety and health officer, or technical authority.

It is important that all hazards are identified and an assessment is made of the potential risk from the hazard. This allows for the determination of the real danger. If a high degree of risk and danger exists, then efforts must be undertaken to alleviate or mitigate them.

Summary

It is important to identify the existing or potential safety hazards and take steps to remove or limit their effects on the workforce. This can be accomplished by utilizing many approaches to control, prevent, or remove safety hazards that could cause injury, illness, and death in the workplace.

Once hazards have been identified, assessed, and controlled, the employer and worker representatives should work together to develop training programs for workers, emergency response procedures, and safety and health requirements for contractors. Someone needs to be responsible for monitoring these activities to ensure they are effective.

The employer is responsible for ensuring that workplace hazards are identified, assessed, and appropriately controlled. Workers must be told about the hazards they face and taught how to control them.

The employer is expected to consult and involve the occupational safety health professionals or worker representatives in the hazard control process. Helping the employer identify, assess, and control hazards is one of the most important roles of the responsible party in the internal responsibility system. Hazards are broadly divided into two groups: hazards that cause illness (health hazards) and those that cause injury (safety hazards). Hazards can be identified by asking what harm could result if a dangerous tool, process, machine, piece of equipment, and so forth failed. Safety and health hazards can be controlled at the source, along the path, or at the level of the worker. Once controls are in place, they must be checked periodically to make sure they are still working properly. Someone should be responsible for auditing the hazard controls in the internal responsibility system and help the employer keep them effective.

Further Readings

Reese, C.D. *Accident/Incident Prevention Techniques (Second Edition)*. Boca Raton, FL: CRC Press, 2012.

Reese, C.D. *Occupational Health and Safety Management (Third Edition)*. Boca Raton, FL: CRC Press, 2016.

Reese, C.D. *Office Building Safety and Health*. Boca Raton, FL: CRC/Lewis Publishers, 2004.

Reese, C.D. and J.V. Eidson. *Handbook of OSHA Construction Safety & Health (Second Edition)*. Boca Raton, FL: CRC/Lewis Publishers, 2006.

Saskatchewan Labour. *Identifying and Assessing Safety Hazards*. http://www.labour.gov.sk.ca/safety, 2007.

United States Department of Labor. *Training Requirements in OSHA Standards and Training Guidelines (OSHA 2254)*. Washington, DC: US Department of Labor, 1998.

United States Department of Labor, Occupational Safety and Health Administration, Office of Training and Education. *OSHA Voluntary Compliance Outreach Program: Instructors Reference Manual*. Des Plaines, IL: US Department of Labor, 1993.

United States Department of Labor. Occupational Safety and Health Administration. *Field Inspection Reference Manual (FIRM): OSHA Instruction CPL 2.103*. Washington, DC: US Department of Labor, September 26, 1994.

United States Department of Labor. Occupational Safety and Health Administration. Office of Training and Education. *OSHA Voluntary Compliance Outreach Program: Instructors Reference Manual*. Des Plaines, IL: US Department of Labor, 1993.

United States Department of Labor, Mine Health and Safety Administration. *Hazard Recognition and Avoidance: Training Manual*. MSHA 0105. Washington, DC: US Department of Labor, Revised May 1996.

41

Personal Protective Equipment

Workers are often faced with hazards that cannot be removed, mitigated, or controlled in some fashion. When this transpires, then steps must be taken to reduce risk, dangers, and hazards. The remaining solution can often be the one of last resort. This why personal protective equipment (PPE) is often used to decrease exposures and provide a semblance of protection and prevention for workers. PPE is not a new concept since suits of armor were used by warriors to prevent injury and death and animal bladders were used by miners exposed to mining products and dust.

Each occupation has its unique hazards. Many hazards faced by various workers cannot be removed, substituted for, isolated, changed to be made safer, or controlled by engineering or mechanical means. Thus, in order to protect workers from exposures to the errant release of energy sources, dangerous materials, extreme natural forces, and unexpected hazards, as a last resort, PPE is utilized in an attempt to offer protection to workers from the hazards they face.

The reason why PPE is the last line of defense is that it must be used to be effective. At times, it provides limited protection, not a guarantee of effectiveness but the best available. Many workers tell stories about the failure of PPE. In most cases, this is not PPE failure but failure to use PPE properly, using improper PPE not meant for the hazard, failure to train workers, and not developing a PPE program for the hazards involved in the workplace.

It has been seen throughout history that certain occupations need to use PPE. The most obvious were warriors and soldiers, from the knights of old to the modern soldier who employs lighter/more appropriate PPE today and the adaptation of such armor for law enforcement employees. Over many years, strides have been made to develop PPE that will better protect workers.

Another reason why PPE has seen use and improvement over time is that the Occupational Safety and Health Administration (OSHA), the Mine Safety and Health Administration (MSHA), and the National Institute for Occupational Safety and Health (NIOSH) have studied and determined the benefits of using it. These efforts have resulted in the codifying of requirements for the use of PPE.

When employees must be present and engineering or administrative controls are not feasible, it is essential to use PPE as an interim control and not a final solution. For example, safety glasses may be required in the work area. Too often, PPE usage is considered the last thing to do in the scheme of hazard control. PPE can provide added protection to the employee even when the hazard is being controlled by other means. There are drawbacks to the use of PPE:

- Hazard still looms.
- Some degree of risk from hazard still exists.
- Protection is dependent upon the worker using PPE.
- PPE may interfere with task performance and productivity.
- It requires supervision.
- It is an ongoing expense.

Many forms of PPE need to be addressed and required when a hazard assessment determines that PPE is the only option left for protecting the workforce. Another reason for PPE is that OSHA has regulations that must be followed, including the following for general industry:

- Eye and face protection (29 CFR 1910.133)
- Respiratory protection (29 CFR 1910.134)
- Head protection (29 CFR 1910.135)
- Foot and leg protection (29 CFR 1910.136)
- Electrical protective equipment (29 CFR 1910.137)
- Hand protection (29 CFR 1910.138)
- Respiratory protection for tuberculosis (29 CFR 1910.139)
- Occupational noise exposure (29 CFR 1910.95)

The following are regulations for PPE and lifesaving equipment for construction:

- Criteria for PPE (29 CFR 1926.95)
- Occupational foot protection (29 CFR 1926.96)
- Protective clothing for fire brigades (29 CFR 1926.97)
- Respiratory protection for fire brigades (29 CFR 1926.98)
- Head protection (29 CFR 1926.100)
- Hearing protection (29 CFR 1926.101)
- Eye and face protection (29 CFR 1926.102)
- Respiratory protection (29 CFR 1926.103)
- Safety belts, lifelines, and lanyards (29 CFR 1926.104)
- Safety nets (29 CFR 1926.105)
- Working over or near water (29 CFR 1926.106)

NOTE: No specific OSHA regulation exists for hand protection for construction.

Any other types of specialized protective equipment needed would be identified as part of the hazard assessment. PPE most often protects from crushing, impact, chemicals, contact, cuts, skin exposure, falling, or inhalation exposure. Such equipment might include body protection for hazardous materials, protective equipment for material handling, protection for welding activities, or protection from exposure to biological, radioactive, or chemical agents.

Because each industry and its occupations (workers) may require unique or specialized PPE because of the hazards identified, workplaces and employers must conduct their own hazard assessment. Some examples, found below, support this statement.

Occupations	Personal Protective Equipment
Electrical	Nonconductive gloves, boots, sleeves, helmet
Logger	Kevlar chaps, mesh face shield, hard toes, hard hat, hearing protection
Medical	Germ-protective clothes, gloves, head covering, respirators, footwear
Firefighters or refinery workers	Flame-retardant clothing
Welders	Leather aprons and gloves, filter-shaded helmets or goggles
Meatcutters	Cut-resistant gloves or metal mesh gloves
Laser workers	Laser goggles
Working at heights	Safety harnesses, lanyards, lifelines
Working over or around water	Life jackets

Why a Hazard Assessment?

The reason for a hazard assessment is to find and evaluate the potential hazards to workers in their workplace. This allows the determination of the need for specific PPE that will add more protection from hazards that cannot be fully controlled by engineering or other means. The hazard assessment formalizes the approach for the selection of the most appropriate personal protection for a particular workplace. The most complex part of PPE selection is, by far, the identifying of chemical protective equipment and the selection of appropriate respirators for existing hazards. This is why an industrial hygienist is often employed.

Why Establish a PPE Program?

A PPE program sets out procedures for selecting, providing, and using PPE as part of an organization's routine operation. A written PPE program, although not mandatory, is easier to establish and maintain than a company policy and easier to evaluate than an unwritten one. To develop a written program, consideration should including the following elements or information:

1. Identify steps taken to assess potential hazards in every employee's work space and in workplace operating procedures.
2. Identify appropriate PPE selection criteria.
3. Identify how you will train employees on the use of PPE, including the following:
   a. What PPE is necessary?
   b. When is PPE necessary?

    c. How to properly inspect PPE for wear and damage.

    d. How to properly put on and adjust the fit of PPE.

    e. How to properly take off PPE.

    f. The limitations of the PPE.

    g. How to properly care for and store PPE.

4. Identify how you will assess employee understanding of PPE training.

5. Identify how you will enforce proper PPE use.

6. Identify how you will provide for any required medical examinations.

7. Identify how and when to evaluate the PPE program.

A personal protective program formalizes the use of PPE as well as sets up standards that will better protect the workforce and criteria for enforcement that are applicable to all employees, including management.

Finally, use PPE for potentially dangerous conditions. Use gloves, aprons, and goggles to avoid acid splashing. Wear earplugs for protection from high noise levels, and wear respirators to protect against toxic chemicals. The use of PPE should be the last consideration in eliminating or reducing the hazards the employee is subjected to because PPE can be heavy, awkward, uncomfortable, and expensive to maintain. Therefore, try to engineer the identified hazards out of the job.

Summary

Many factors need to be considered when selecting PPE to protect employees from workplace hazards. With all of the types of operations that can present hazards and all of the types of PPE available to protect the different parts of a worker's body from specific types of hazards, this selection process can be confusing and at times overwhelming. Because of this, OSHA requires that employers implement a PPE program to help employers systematically assess the hazards in the workplace and select the appropriate PPE that will protect workers from those hazards. As to why PPE should be a part of the occupational safety and health (OSH) program, the following are some reasons:

- To assess the workplace to identify equipment, operations, chemicals, and other workplace components that could harm employees
- To guide the implementation of engineering controls and work practices to control or eliminate these hazards to the extent feasible
- To facilitate the selection the appropriate types of PPE to protect your employees from hazards that cannot be eliminated or controlled through engineering controls and work practices
- To inform employees why the PPE is necessary and when it must be worn
- To train employees on how to use and care for the selected PPE and how to recognize PPE deterioration and failure
- To require employees to wear the selected PPE in the workplace

The basic information presented here attempts to establish and illustrate a logical, structured approach to hazard assessment and PPE selection and use. These steps must be followed in order to prevent occupational injuries and illnesses.

Further Readings

Reese, C.D. *Accident/Incident Prevention Techniques (Second Edition)*. Boca Raton, FL: CRC Press, 2012.

Reese, C.D. *Occupational Health and Safety Management: A Practical Approach (Third Edition)*. Boca Raton, FL: CRC Press, 2015.

Reese, C.D. and J.V. Eidson. *Handbook of OSHA Construction Safety & Health (Second Edition)*. Boca Raton, FL: CRC Press, 2006.

42

Occupational Safety and Health Administration (OSHA)

Prior to 1970, it became apparent that the proliferation of injuries, illnesses, and deaths in the workplace was not being taken serious by employers as well as others. This basically forced the federal government to act. This is why in 1970, Congress enacted the Occupational Safety and Health Act (OSHAct) to protect workers from workplace hazards. This law requires companies and businesses to develop workplaces free from hazards with the intention of protecting the American workforce. A product of the OSHAct was the Occupational Safety and Health Administration (OSHA), a federal agency that has approximately 2,200 inspectors to protect the safety and health of 130 million employees. Following the OSHA standards leads to safer workplaces and has improved productivity and profit for US industries.

As an employer, this why it is critical to have an understanding of how OSHA works, achieves its mission, and strives to protect the American workforce. OSHA can be an ally or a thorn in a company's side, depending upon its approach to job safety and health.

Workers should expect to go to work each day and return home uninjured and in good health. There is no logical reason that a worker should be part of workplace carnage. Workers do not have to become one of the yearly workplace statistics.

Employers who enforce the occupational safety and health (OSH) rules and safe work procedures are less likely to have themselves or their workers become one of the 6,500 occupational trauma deaths, one of the 90,000 occupational illness deaths, or even one of the 6.8 million nonfatal occupational injuries and illnesses that occur each year in the United States.

OSHA and its regulations should not be the driving force that ensures workplace safety and health. Since OSHA has limited resources and inspectors, enforcement is usually based on serious complaints, catastrophic events, and workplace fatalities. The essence of workplace safety and the strongest driving catalyst should first be the protection of the workforce, followed by economic incentives for the employer. Employers having a good safety and health program and record will reap the benefits: a better opportunity to win more customers; lower insurance premiums for workers' compensation; decreased liability; and increased employee morale and efficiency. Usually, safety and health are linked to the bottom line (company's income), which is seldom perceived as humanitarian.

During the many years preceding OSHA, it became apparent that employers needed guidance and incentives to insure safety and health on the jobsite. The employer needed to realize that workers had a reasonable right to expect a safe and healthy workplace. This guidance and the guarantee of a safe and healthy workplace came to fruition with the enactment of the OSHAct. The following is why the OSHA was created by the Act:

- To encourage employers and employees to reduce workplace hazards and to improve existing safety and health programs or implement new programs
- To provide for research in OSH in order to develop innovative ways of dealing with OSH problems

- To establish "separate but dependent" responsibilities and rights for employers and employees for the achievement of better safety and health conditions
- To maintain a reporting and record-keeping system to monitor job-related injuries and illnesses
- To establish training programs to increase the numbers and competence of OSH personnel
- To develop mandatory job safety and health standards and enforce them effectively
- To provide for the development, analysis, evaluation, and approval of state OSH programs

Thus, the purpose of OSHA is to insure, as much as possible, a healthy and safe workplace free of hazardous conditions for workers in the United States.

OSHA Standards

OSHA standards found in the Code of Federal Regulations (CFR) include the standards for the following industry groups: construction; maritime; agriculture; general industry, which includes manufacturing; transportation and public utilities; wholesale and retail trades; finance; insurance; and services. OSHA standards and regulations for OSH are found in Title 29 of the CFR and can be obtained through the Government Printing Office (GPO). The standards for specific industries are found in Title 29 of the CFR.

Protections under the OSHAct

Usually, all employers and their employees are considered to be protected under the OSHAct, with the exception of the following:

- Self-employed persons
- Farms where only immediate family members are employed
- Workplaces already protected under federal statutes by other federal agencies such as the Department of Energy and the Mine Safety and Health Administration
- State and local employees

National Institute for Occupational Safety and Health (NIOSH)

Although the formation of National Institute for Occupational Safety and Health (NIOSH) was a requirement of the OSHAct of 1970, NIOSH is not part of OSHA. NIOSH is one of the Centers for Disease Control and Prevention, headquartered in Atlanta, Georgia. NIOSH reports to the Department of Health and Human Services (DHHS) and not to the Department of Labor (DOL) as OSHA does. Its functions are as follows:

- To recommend new safety and health standards to OSHA
- To conduct research on various safety and health problems
- To conduct health hazard evaluations (HHEs) of the workplace when called upon
- To publish an annual listing of all known toxic substances and recommended exposure limits (RELs)
- To conduct training that will provide qualified personnel under the OSHAct

An employer, worker's representative, or worker can request a HHE from NIOSH to have a potential health problem investigated. It is best to use the NIOSH standard form. It can be obtained by calling 1-800-35-NIOSH.

Occupational Safety and Health Review Commission (OSHRC)

The Occupational Safety and Health Review Commission (OSHRC) was established, under the OSHAct, to conduct hearings when OSHA citations and penalties are contested by employers or by their employees. As with NIOSH, OSHRC formation was a requirement of the OSHAct, but it is a separate entity from OSHA.

Employer Responsibilities under the OSHAct

The employer is held accountable and responsible under the OSHAct. The General Duty Clause, Section 5(a)(1) of the OSHAct, states that employers are obligated to provide a workplace free of recognized hazards that are likely to cause death or serious physical harm to employees. This is why employers must do the following:

- Abide and comply with the OSHA standards
- Maintain records of all occupational injuries and illnesses
- Maintain records of workers' exposure to toxic materials and harmful physical agents
- Make workers aware of their rights under the OSHAct
- Provide, at a convenient location and at no cost, medical examinations to workers when the OSHA standards require them
- Report within 8 hours to the nearest OSHA office all occupational fatalities or catastrophes where three or more employees are hospitalized
- Abate cited violations of the OSHA standard within the prescribed time period
- Provide training on hazardous materials and make Safety Data Sheets (SDSs) that provide information on hazards available to worker upon request

- Assure that workers are adequately trained under the regulations
- Post information required by OSHA such as citations, hazard warnings, and injury/illness records

Workers' Rights and Responsibilities under the OSHAct

Workers have many rights under the OSHAct. These rights include the right to do the following:

- Review copies of appropriate standards, rules, regulations, and requirements that the employer should have available at the workplace
- Request information from the employer on safety and health hazards in the workplace, precautions that may be taken, and procedures to be followed if an employee is involved in an accident or is exposed to toxic substances
- Access relevant worker exposure and medical records
- Be provided personal protective equipment (PPE)
- File a complaint with OSHA regarding unsafe or unhealthy workplace conditions and request an inspection
- Not be identified to the employer as the source of the complaint
- Not be discharged or discriminated against in any manner for exercising rights under the OSHAct related to safety and health
- Have an authorized employee representative accompany the OSHA inspector and point out hazards
- Observe the monitoring and measuring of hazardous materials and see the results of the sampling, as specified under the OSHAct and as required by OSHA standards
- Review the occupational injury and illness records (OSHA 200 or equivalent) at a reasonable time and in a reasonable manner
- Have safety and health standards established and enforced by law
- Submit to NIOSH a request for an HHE of the workplace
- Be advised of OSHA actions regarding a complaint and request an informal review of any decision not to inspect or issue a citation
- Participate in the development of standards
- Speak with the OSHA inspector regarding hazards and violations during the inspection
- File a complaint and receive a copy of any citations issued and the time allotted for abatement
- Be notified by the employer if the employer applies for a variance from an OSHA standard and testify at a variance hearing and appeal the final decision
- Be notified if the employer intends to contest a citation, abatement period, or penalty
- File a Notice of Contest with OSHA if the time period granted to the company for correcting the violation is unreasonable, provided it is contested within 15 working days of the employer's notice

- Participate at any hearing before the OSHA Review Commission or at any informal meeting with OSHA when the employer or a worker has contested an abatement date
- Appeal the OSHRC's decisions in the US court of appeals
- Obtain a copy of the OSHA file on a facility or workplace

Along with rights go responsibilities, and workers should be expected to conform to these responsibilities. Workers are expected to do the following:

- Comply with the OSHA regulations and standards
- Not remove, displace, or interfere with the use of any safeguards
- Comply with the employer's safety and health rules and regulations
- Report any hazardous conditions to the supervisor or employer
- Report any job-related injuries and illnesses to the supervisor or employer
- Cooperate with the OSHA inspector during inspections when requested to do so

One point that should be kept in mind is that it is the employer's responsibility to assure that employees comply with OSHA regulations and safety and health rules. Workers are not held financially accountable by OSHA for violations of OSHA regulations. It is entirely up to the employer to hold employees accountable. With the accountability and responsibility falling upon the employer, he/she must take control and direct the safety and health effort at their workplace.

Discrimination against Workers

Workers have the right to expect safety and health on the job without fear of punishment. This is spelled out in Section 11(c) of the OSHAct and under 49 USC 31105 (formerly Section 405) for the trucking industry. The law states that employers shall not punish or discriminate against workers for exercising rights such as the following:

- Complaining to an employer, union, or OSHA (or other government agency) about job safety and health
- Filing a safety and health grievance
- Participating in OSHA inspections, conferences, and hearings, or OSHA-related safety and health activities

Right to Information

Workers have a "right to know." This means that the employer must establish a written, comprehensive hazard communication program that includes provisions for container labeling, MSDSs, and an employee training program. The program must include the following:

- A list of the hazardous chemicals in the workplace
- The means the employer uses to inform employees of the hazards of nonroutine tasks
- The way the employer will inform other employers of the hazards to which his/ her employees may be exposed

Workers have the right to information regarding the hazards to which they are or will be exposed. They have the right to review plans such as the hazard communication plan. They have a right to see a copy of an MSDS during their shift and receive a copy of it when requested. Also, information on hazards that may be brought to the workplace by another employer should be available to workers. Other forms of information such as exposure records, medical records, etc., are to be made available to workers upon request.

OSHA Inspections

OSHA has the right to conduct workplace inspections as part of its enforcement mandate. OSHA can routinely initiate an unannounced inspection of a business. Other inspections occur due to fatalities and catastrophes, as part of routine program inspections, or due to referrals and complaints. These occur during normal working hours.

Workers have the right to request an inspection. The request should be in writing (either by letter or by using the OSHA complaint form to identify the employer and the alleged violations). Send the letter or form to the area director or state OSHA director. If workers receive no response, they should contact the OSHA regional administrator. It is beneficial to call the OSHA office to verify its normal operating procedures. If workers allege an imminent danger, they should call the nearest OSHA office.

These inspections include the following: checking company records; reviewing compliance with the hazard communication standard, fire protection, and PPE; and review of the company's safety and health plan. This inspection will include conditions, structures, equipment, machinery, materials, chemicals, procedures, and processes. These inspections may result in discovery of violation, which can lead to monetary citations.

Worker Training

Many standards promulgated by OSHA specifically require the employer to train employees in the safety and health aspects of their jobs. Other OSHA standards make it the employer's responsibility to limit certain job assignments to employees who are "certified," "competent," or "qualified"—meaning that employees have had special previous training, in or out of the workplace. OSHA regulations imply that an employer has assured that a worker has been trained prior to being designated as the individual to perform a certain task.

In order to make a complete determination of the OSHA requirement for training, one would have to go directly to the regulation that applies to the specific type of activity. The regulation may mandate hazard training, task training, and length of the training, as well as specifics to be covered by the training.

It is always a good idea for the employer, as well as the worker, to keep records of training. These records may be used by a compliance inspector during an inspection, after an accident resulting in injury or illness, as proof of good intentions by the employer or compliance with training requirements for workers, including new workers and those assigned new tasks.

Occupational Injuries and Illnesses

The recording and reporting of occupational injuries and illness requirements can be found in 29 CFR 1904—Recording and Reporting Occupational Injuries and Illnesses. This regulation has been revised and came into effect as of January 2002. These requirements are summarized in the following paragraphs.

Any illness that has been caused by exposure to environmental factors such as inhalation, absorption, ingestion, or direct contact with toxic substances or harmful agents and has resulted in an abnormal condition or disorder that is acute or chronic is classified as an occupational disease. Repetitive-motion injuries are also included in this category. All illnesses are recordable, regardless of severity. Injuries are recordable under the following conditions:

- An on-the-job death occurs regardless of length of time between injury and death.
- One or more lost workdays occurs.
- Restriction of work or motion transpires.
- Loss of consciousness occurs.
- Worker is transferred to another job.
- Worker receives medical treatment beyond first aid.

Employers with more than 10 employees are required to complete and maintain occupational injury and illness records. The OSHA 301, Injury and Illness Incident Report, or equivalent, must be completed within 7 days of the occurrence of an injury at the work site and must be retained for 5 years. Also, the OSHA 300, Log of Work-Related Injuries and Illnesses, is to be completed within 7 days when a recordable injury or illness occurs, and maintained for 5 years. The OSHA 300A, Summary of Work-Related Injuries and Illnesses, must be posted yearly from February 1 to April 30. OSHA forms can now be maintained on the computer until they are needed.

Summary

It is envisioned that this is why OSHA is an asset to employers. Knowledge has been shown to fix accountability, as well as responsibility, in those who claim ignorance of it.

The workplace is where both labor and management spend the bulk of their waking hours. With this in mind, the safety and health of those in the workplace should be everyone's concern and responsibility.

Employers and safety and health professionals need to know how OSHA provides for worker safety and health on work sites. This will also assist in assuring that the workers' rights are protected and give them the knowledge to help mitigate safety and health issues and problems that may arise. This type of knowledge should ensure a safer and more productive work site. Respect for the efficient, effective, and proper use of the safety and health rules will have a positive effect upon those in the workplace.

Although it is the ultimate responsibility of the employer to provide for workplace safety and health, adherence to OSHA occupational safety and health rules is the foundation upon which a good safety and health program can be built. The program should hold everyone responsible for the well-being of those in the workplace, including the employer, managers, supervisors, and workers. All should abide by the safety and health rules and the OSHA standards. Together, and through cooperation, all parties can assure a safe and healthy workplace. A safe and healthy home away from home is, and should, be the ultimate goal.

Further Readings

Anton, T.J. *Occupational Safety and Health Management (Second Edition)*. New York: McGraw Hill, Inc., 1989.

Bertinuson, J. and S. Weinstein. *Occupational Hazards of Construction: A Manual for Building Trades' Apprentices*. Labor Occupational Health Program, University of California, Berkeley, CA, 1978.

Blosser, F. *Primer on Occupational Safety and Health*. Washington, DC: The Bureau of National Affairs, 1992.

Murphy, W.C. and J.R. Hanson. *A Maine Guide to Employment Law*. Orono, ME: The University of Maine, 1995.

National Safety Council. *Protecting Workers Lives: A Safety and Health Guide for Unions (Second Edition)*. Itasca, IL: National Safety Council, 1992.

Reese, C.D. *Accident/Incident Prevention Techniques (Second Edition)*. Boca Raton, FL: CRC Press, 2012.

Reese, C.D. *Occupational Health and Safety Management (Third Edition)*. Boca Raton, FL: CRC Press, 2016.

Reese, C.D. and J.V. Eidson. *Handbook of OSHA Construction Safety & Health (Second Edition)*. Boca Raton, FL: CRC/Lewis Publishers, 2006.

United States Department of Labor/OSHA. *Access to Medical Records and Exposure Records (OSHA 3110)*. Washington, DC: US Department of Labor, 1988.

United States Department of Labor/OSHA. *All About OSHA (OSHA 2056)*. Washington, DC: US Department of Labor, 1985.

United States Department of Labor/OSHA. *OSHA: Employee Workplace Rights (OSHA 3021)*. Washington, DC: US Departrment of Labor, 1991.

United States Department of Labor/OSHA. *Recordkeeping Guidelines for Occupational Injuries and Illnesses (OMB No. 1220-0029)*. Washington, DC: US Departrment of Labor, 1986.

United States Department of Labor/OSHA. *Training Course in OSHA for the Construction Industry (Course #500)*. OSHA Training Institute, Des Plaines, IL: US Department of Labor, 1997.

Section I

Accident/Incident Prevention Techniques

Accident prevention techniques are a group of tried-and-tested applications for prevention that have been used for ages, it seems. Although many alterations to these approaches have been used, their basic premises continue to be used to this day and have proven to be useful to those managing occupational safety and health (OSH).

The contents of this section are as follows:

Chapter 43—Safe Operating Procedures

Chapter 44—Job Safety Analysis

Chapter 45—Job Safety Observation

Chapter 46—Fleet Safety

Chapter 47—Preventive Maintenance Programs

43

Safe Operating Procedures

Why are safe operating procedures (SOPs) or standard operating procedures necessary? SOPs or standard operating procedures should be included as the product of a job safety analysis or job hazard analysis program of a company. The SOP is a set of written instructions that document a routine or repetitive activity followed by an organization. The development and use of standard or safe operating procedures is an integral part of a successful quality system as it provides workers with information to perform a job properly and safely while facilitating consistency in the quality and integrity of a product or end result. SOPs provide precise directions for performing a job, operating a piece of equipment, or driving a vehicle in a safe and healthy manner, thus preventing an accident or incident. Thus, SOPs are an important component of a prevention undertaking.

Workers may not automatically understand a task just because they have experience or training. Thus, many jobs, tasks, and operations are best supported by an SOP. The SOP walks the worker through the steps of how to do a task or procedure in a safe manner and calls attention to the potential hazards at each step. The development and use of SOPs minimizes variation and promotes quality through consistent implementation of a process or procedure within the company, in the event that there are temporary or permanent personnel changes. It minimizes opportunities for miscommunication and can address safety and health concerns. SOPs are frequently used as a checklist by inspectors when auditing procedures. Ultimately, the benefits of a valid SOP are reduced work effort and safe performance, along with improved comparability, credibility, and legal defensibility.

You might ask why an SOP is needed if the worker has already been trained to do the job or task. As one might remember from the job safety analysis, it usually keys in on those particular jobs that pose the greatest risk of injury or death. This is why an SOP takes the guesswork out of operating or performing a particular job or task in a safe manner. These are high-risk types of work activities and definitely merit the development and use of an SOP. There are times when an SOP, or step-by-step checklist, is useful. Some of these times are as follows:

1. A new worker is performing a job or task for the first time.
2. An experienced worker is performing a job or task for the first time.
3. A worker has not performed the job in a long time.
4. An experienced worker is performing a job, which he/she has not done recently.
5. Mistakes could cause damage to equipment or property.
6. A job is done on an intermittent or infrequent basis.
7. A new piece of equipment or different model of equipment is obtained.

8. Supervisors need to understand safe operation to be able to evaluate performance.

9. A procedure or action within an organization is repetitive and is carried out in the same way each time.

10. A procedure is critically important, no matter how seldom performed, and must be carried out exactly according to detailed, stepwise instructions.

11. There is a need to standardize the way a procedure is carried out to ensure quality control or system compatibility.

When airline pilots fly, the most critical parts of the job are takeoffs and landings. Since these are such crucial aspects of flying, a checklist for proceeding in a safe manner is used to mitigate the potential for mistakes. It is vital to provide help when a chance for error can result in grave consequences. Similarly, laminated SOPs should be placed on equipment, machines, and vehicles for individuals who need a refresher prior to operation because they have not used the equipment or have performed a task on an infrequent basis.

Few people or workers want to admit that they do not know how to perform a job or task. They will not ask questions, let alone ask for help in doing an assigned task. This is the time when a laminated SOP or checklist could be placed at the work site or attached to a piece of equipment. This can prove to be a very effective accident prevention technique. It can safely walk a worker through the correct sequence of necessary steps and thus avoid the exposure to hazards, which can put the worker at risk of injury, illness, or death.

These SOPs could be used when helicopters are employed for lifting, industrial forklifts are utilized, materials are moved manually, etc. They should list the sequential steps required in order to perform the job or task safely, the potential hazards involved, and the personal protective equipment needed. Each step in the SOP should provide all the information required to accomplish the task in a safe manner.

If there is no annual training, then the use of SOPs may instill confidence in the workers that they have a way of refreshing their memories on how to do their tasks. When changing procedures, the SOP should be updated to assure that the changed procedure is performed safely. A checklist is one form of SOP. The checklist is very effective when you are trying make sure that every step is being accomplished.

SOPs are only useful when they are up to date and readily accessible at the actual job or task site. Since we have the ability to store SOPs in the computer, revisions can be made simply and are therefore easily changed if the company receives good suggestions from supervisors or workers. An SOP is only one accident prevention technique or component of any safety and health initiative. There are specific jobs or tasks that lend themselves well to this approach. Make sure that SOPs are used when they benefit the type of work being performed the most and not as a cure-all for all accidents and injuries. Use them as one of the many tools for accident prevention.

SOPs or standard operating procedures given to a work process are created and written after identifying and assessing the safety and health risks involved in a work activity, and they identify how to eliminate the risks or, alternatively, what reasonably practicable actions can be done to minimize those risks. Simply stated, an SOP explains how to do a certain task and remain safe and healthy.

Why Are SOPs Beneficial?

SOPS must provide a benefit to the safety and health program, the employer, and the supervisors and workers that make use of them. Some of these benefits are as follows:

- SOPs reduce the risk of injury and illness.
- They provide a stable reference for doing the job in a safe and healthy manner.
- They improve the awareness and understanding of risks in the workplace and how to handle them.
- They can be generated using consultation with workers.
- They are a tested and agreed-upon procedure that can be used to train workers in the steps to use to perform an activity in a safe and healthy manner.
- They are a standard or reliable reference point of conducting observation and inspection.
- SOPs are indicators to Occupational Safety and Health Administration (OSHA) inspectors that safety and health is a point of emphasis in the company.
- SOPs indicates a company places emphasis on providing a safe and healthy workplace.

Other benefits include that they reduce the learning curve/training time for new employees, ensure continuity if another worker must take over the activity, and assure standardization in performing a task while insuring a standard product or task.

What Are the Components Needed to Develop an SOP?

SOPs are intended to provide clear instructions for safely conducting activities involved in each procedure covered. SOPs help to assure that consistent work activities are using the appropriate manufacturer's guidelines and instructions, and they give other pertinent safety and health information from other resources. They also allow individuals with specific safety knowledge and expertise to address at least the following elements:

1. Steps for each operating phase:
   a. Initial start-up
   b. Normal operations
   c. Temporary operations
   d. Emergency shutdown including the conditions under which emergency shutdown is required, and the assignment of shutdown responsibility to qualified operators or workers to ensure that emergency shutdown is executed in a safe and timely manner

    e. Emergency operations

    f. Normal shutdown

    g. Start-up following a turnaround, or after an emergency shutdown

2. Operating limits:

    a. Consequences of deviation

    b. Steps required for correcting or avoiding deviations

3. Safety and health considerations:

    a. Properties of, and hazards presented by, the chemicals and materials used in the process and hazards involved in the task

    b. Precautions necessary to prevent exposure, including engineering controls, administrative controls, and personal protective equipment

    c. Control measures to be taken if physical contact or airborne exposure occurs

    d. Quality control for raw materials, control of hazardous chemical inventory levels, and any other special or unique hazards

4. Safety systems and their functions

SOPs should be readily accessible to employees who work with or maintain a process or operation. The SOPs should be reviewed as often as necessary to assure that they reflect current operating practice, including changes that result from changes in the process or procedures due to new understanding or requirements of technology, equipment, and facilities. The employer should develop and implement safe work practices to provide for the control of hazards during operations such as equipment/ machine operation, lockout/tagout, confined space entry, and opening process equipment or piping, and control over an entrance into a facility by maintenance, contractor, laboratory, or other support personnel. These safe work practices must apply to both employees and contract employees.

Why SOPs Are Poorly Written

SOPs are often poorly written because little thought is given or effort made to get them right. At times, they are mandated as a quick fix for a perceived problem. An organized and thoughtful approach will yield SOPs that are truly useable by the workers.

SOPs should be written in a concise, step-by-step, easy-to-read format. The information presented should be unambiguous and not overly complicated. The active voice and present verb tense should be used. The word "you" should not be used, but implied. The document should not be wordy, redundant, or overly lengthy. Keep it simple and short. Information should be conveyed clearly and explicitly to remove any doubt as to what is required. Also, use a flowchart to illustrate the process being described.

The most common problems and errors found in SOPs are summarized in the following list.

1. Enumerating responsibilities for carrying out a procedure rather than stating who does the procedure and how. Regulations are the place for delineating responsibilities, not SOPs.

2. Failure to clearly state who carries out which step in the procedure. The *who* is as important as the *what*.

3. Inclusion of steps or procedures performed by persons outside the organization. This information has no place in an SOP because it involves actions that are beyond the direct control of the organization. Include only those steps that are carried out by the employees in the immediate organization; all else is irrelevant.

4. Vagueness and imprecision. What if the reader of the SOP cannot figure out exactly who (job description) is required to carry out a step in the procedure, and furthermore cannot determine precisely how it is to be carried out? Obviously, then, the SOP has failed in its primary objective, communication. This is why a prime function of the reviewer is to check to see if the writer has conveyed the message clearly and unequivocally.

Why SOPs Work

Safe or standard operating procedures describe tasks to be performed, data to be recorded, operating conditions to be maintained, samples to be collected, and safety and health precautions to be taken. The procedures need to be technically accurate, understandable to employees, and revised periodically to ensure that they reflect current operations. Operating procedures should be reviewed by engineering staff and operating personnel to ensure that they are accurate and provide practical instructions on how to actually carry out job duties safely.

Operating procedures include specific instructions or details on what steps are to be taken or followed in carrying out the stated procedures. These instructions for each operating procedure should include the applicable safety precautions and should contain appropriate information on safety implications. For example, operating procedures should address operating parameters and contain operating instructions about pressure limits, temperature ranges, flow rates, what to do when an upset condition occurs, what alarms and instruments are pertinent if an upset condition occurs, and other subjects. For example, operating instructions to properly drive a powered vehicle during start-up or shutdown involve unique processes. In some cases, different parameters are required from those of normal operation. These operating instructions need to clearly indicate the distinctions between normal and deviant operations, such as instructions and parameters on the appropriate allowances for driving while fully loaded.

Operating procedures and instructions are important for training operating personnel. Operating procedures are the standard operating practices for operations. Operators and operating staff, in general, need to have a full understanding of operating procedures. If workers are not fluent in English, then procedures and instructions need to be prepared in a second language understood by those workers. In addition, operating procedures need to be changed when there is a change in the process as a result of a management decision or procedural changes. The consequences of operating procedure changes need to be fully evaluated and communicated to personnel. For example, mechanical changes to a process made by the maintenance department (like changing a valve from steel to brass or other subtle changes) need to be evaluated to determine if operating procedures and practices also need to be changed.

All management changes and actions must be coordinated and integrated with current operating procedures, and operating personnel must be oriented to the changes in procedures before the change is made. When the process is shut down in order to make the changes, the operating procedures must be updated before start-up of the process.

Training in how to handle upset conditions must be accomplished as well as training on what operating personnel are to do in emergencies, such as when a pump seal fails or a pipeline ruptures. Communication must also be maintained between operating personnel and workers performing work within the operating or production area, especially with nonroutine tasks. Hazards of the task are to be conveyed to both operating personnel (e.g., supervisors), in accordance with established procedures, and those workers performing the actual tasks.

Summary

A standard or safe operating procedure is a document that describes the regularly recurring operation relevant to quality and safe performance. The purpose of an SOP is to carry out the operations correctly and always in the same manner. An SOP should be available at the place where the work is done. SOPs are organizational tools that provide a foundation for training new employees, for refreshing the memories of management and experienced employees, and for ensuring that important procedures are carried out in a standard specified way. The principal function of an SOP is to provide detailed, step-by-step guidance to employees who are required to carry out a certain procedure. In this instance, it serves not only as a training aid but also as a means of helping to ensure that the procedure is carried out in a standard, approved manner.

Another important function of an SOP is to keep management informed about the way functions are performed in areas under their supervision. A complete file of well-written, up-to-date SOPs is an indication of good management and provides management with instant access to information on functional details of the organization for which they are responsible. This is of enormous benefit during inspections and management reviews, to say nothing of providing timely answers to unanticipated questions from superiors. An SOP is a compulsory instruction. If deviations from this instruction are allowed, the conditions for these should be documented, including who can give permission for this and what exactly the complete procedure will be. The original should rest in a secure place, while working copies should be authenticated with stamps and/or signatures of authorized persons.

The overridingly important feature of a good SOP is that it communicates what is to be done in a clear, concise, and stepwise manner. The most important person to whom it must communicate is typically the new employee, who may have little or no experience with the procedure in question. Therefore, it is imperative that the writer of an SOP figuratively places himself/herself in the position of a new, inexperienced employee in order to appreciate what must be communicated and how to communicate it.

The content of an SOP should be comprehensive in terms of how to get the procedure accomplished but should not encompass matters not directly relevant because to digress does not directly address the issue of how to get the procedure accomplished and exactly who is to do it.

Perhaps the best advice concerning the content of an SOP is this. Ask yourself the questions who, what, where, when, and how. If the SOP answers all of these questions, it is complete. If not, revise it until it does in as clear and logical an order as possible.

Further Readings

Reese, C.D. *Accident/Incident Prevention Techniques (Second Edition)*. Boca Raton, FL: CRC Press, 2012.

Reese, C.D. *Occupational Health and Safety Management (Third Edition)*. Boca Raton, FL: CRC Press, 2016.

Reese, C.D. and J.V. Eidson. *Handbook of OSHA Construction Safety & Health (Second Edition)*. Boca Raton, FL: CRC/Lewis Publishers, 2006.

United States Department of Labor, Mine Safety and Health Administration, *Accident Prevention (Safety Manual No. 4)*. Beckley, WV: US Department of Labor, Revised 1990.

United States Department of Labor, Mine Safety and Health Administration, *Job Safety Analysis: A Practical Approach (Instruction Guide No. 83)*. Beckley, WV: US Department of Labor, 1990.

United States Department of Labor, Mine Safety and Health Administration, *Job Safety Analysis (Safety Manual No. 5)*. Beckley, WV: US Department of Labor, Revised 1990.

United States Department of Labor, Mine Safety and Health Administration, *Safety Observation (MSHA IG 84)*. Beckley, WV: US Department of Labor, Revised 1991.

United States Department of Labor, Occupational Safety and Health Administration, Office of Training and Education. *OSHA Voluntary Compliance Outreach Program: Instructors Reference Manual*. Des Plaines, IL: US Department of Labor, 1993.

44

Job Safety Analysis

Job safety analysis (JSA), sometimes called job hazard analysis (JHA), is used as the foundational part of an accident prevention program. Using JSAs, many accident prevention tools and process can be devised; for example, job safety observations can be conducted, safe operating procedures can be developed, potential hazards can be identified, and audit instruments can be constructed. JSAs can also be used as a training tool. Also, potential types of accidents can be identified, such as struck-against, struck-by, contact-with, contacted-by, caught-in, caught-on, caught-between, fall-same-level, fall-to-below, overexertion, or exposure.

Why JSA?

The reason to develop JSAs for each job at the company's operations is as follows:

- Establish proper job procedures
- Identify potential or existing job hazards
- Determine the best way to perform or eliminate hazards
- Improve job methods
- Reduce cost
- Reduce absenteeism
- Reduce workers' compensation
- Increase productivity
- Increase hazard reporting
- Determine what personal protective equipment (PPE) is needed for each job
- Develop preventive measures
- Develop new safer job procedures or solutions

The easiest and possibly most effective method is the step-by-step process of the JHA or JSA. The hazard analysis process looks at jobs or processes. Done for every job, a JHA or JSA ensures safe steps, teaches new workers, eliminates or controls hazardous materials, and much more. Some companies have work teams complete JHAs or JSAs on every job or process and then use them as a guide to do the job. The JHA is a hazard identification tool, an analysis tool, a training tool, and an accident prevention tool all rolled into one.

JSA/JHA is a process used to determine hazards and safe procedures for each step of a job. A specific job, or work assignment, can be separated into a series of relatively simple steps. The hazards associated with each step can be identified, and solutions can be developed to control each hazard. A simple form can be used to carry out a JHA.

A JSA/JHA is a procedure that integrates accepted safety and health principles and practices into a specific task or job procedure. This is why in a JSA/JHA, each basic step of the job is to identify potential hazards and to recommend the safest way to do the job. Jobs that should have JSA/JHA conducted on them and receive attention first are as follows:

- Jobs with the highest injury or illness rates
- Jobs with the potential to cause severe or disabling injuries or illnesses, even if there is no history of previous accidents
- Jobs in which one simple human error could lead to a severe accident or injury
- Jobs that are new to the operation or have undergone changes in processes and procedures
- Jobs complex enough to require written instructions

Much of the information within this chapter comes from the United States Department of Labor's Mine Safety and Health Administration material entitled *The Job Safety Analysis Process: A Practical Approach.* The precept behind using JSA/JHA is that fatalities, accidents, and injuries can be reduced by working together and sharing safety and health knowledge. An accident prevention method that has proven effective in industry is the JSA/JHA program.

JSA/JHA is a basic approach to developing improved accident prevention procedures by documenting the firsthand experience of workers and supervisors, and at the same time, it tends to instill acceptance through worker participation. JSA/JHA can be a central element in a safety program, and the most effective safety programs are those that involve employees. Each worker, supervisor, and manager should be prepared to assist in the recognition, evaluation, and control of hazards. Worker participation is important to efficiency, safety, and increased productivity. Through the process of JSA/JHA, these benefits are fully realized. This process can begin in the following ways:

- Involve employees
- Review accident history
- Conduct a preliminary job review
- List, rank, and set priorities for hazardous jobs
- Outline the steps or tasks

JSA/JHA is a process used to determine hazards of, and safe procedures for, each step of a job. A specific job, or work assignment, can be separated into a series of relatively simple steps. The hazards associated with each step can be identified, and solutions can be developed to control each hazard.

Why Perform a JSA/JHA?

The following questions need to be answered in conducting a JSA:

- What can go wrong?
- What are the consequences?

- How could it happen?
- What are the contributing factors?
- How likely is it that the hazard will occur?

The answers to these questions will help in hazard identification. Some of the other accident-specific questions that need to be answered are as follows:

- Can any body part get caught in or between objects?
- Do tools, machines, or equipment present any hazards?
- Can the worker make harmful contact with moving objects?
- Can the worker slip, trip, or fall?
- Can the worker suffer strain from lifting, pushing, or pulling?
- Is excessive noise or vibration a problem?
- Is the worker exposed to extreme heat or cold?
- Is there danger from falling objects?
- Is lighting a problem?
- Can weather conditions affect safety?
- Is harmful radiation a possibility?
- Can contact be made with hot, toxic, or caustic substances?
- Are there dusts, fumes, mists, or vapors in the air?

Why Four Basic Steps in Developing a JSA/JHA?

JSA/JHA involves four basic steps:

1. Select a job to be analyzed.
2. Separate the job into its basic steps.
3. Identify the hazards associated with each step.
4. Control each hazard.

This is why looking at these four steps in detail will help explain the process and value of this type of analysis.

Why Select Jobs by Using Criteria?

The first step of a JSA/JHA is to select a job to be analyzed. The sequence in which jobs are analyzed should be established when starting a JSA/JHA program. Potential jobs for analysis should have sequential steps and a work goal when these steps are performed.

To use the JSA/JHA program effectively, a method must be established to select and prioritize the jobs to be analyzed. The jobs must be ranked in order of greatest accident potential. Jobs with the highest risks should be analyzed first. Workers and supervisors may or may not be involved with the ranking process, but if asked to rank or prioritize jobs to be analyzed, the following criteria should be used: accident frequency, accident severity, judgment and experience, new jobs, nonroutine jobs, job changes, and routine jobs.

A form should be developed that will allow for a written record for each JSA to be maintained. In the JSA process, it is easiest to deal with each column of the form separately. Logically, the job should be broken down into its basic steps first. Each step or activity should briefly describe what is done. Each activity should be listed on the form in the order it is accomplished. At this point in the analysis, two things need to be done: first, observe the job actually being performed (if possible, this should be done by more than one person), and second, involve at least one employee who does the job regularly in the analysis. The first step might be to do a walk-around inspection. When outlining the job steps, it will be tempting to get very detailed and list how to do the job rather than the basic job steps. The list of job steps in the "Sequence of Basic Job Steps" column of the JSA/JHA form will continue to be broken down into manageable steps. It is important that the JSA/JHA accurately describes the work. There may be some disagreement about basic job steps.

Why Identify the Hazards Associated with Each Job Step?

After all basic steps of the operation of a piece of equipment or job procedure have been listed, each job step needs to be examined to identify hazards associated with each job step. The purpose is to identify and list the hazards that could possibly arise in each step of the job. Some hazards are more likely to occur than others, and some are more likely to produce serious injuries than others. Consider all reasonable possibilities when identifying hazards. To make this task manageable work with the basic types of accidents, the question to ask is, "Can any of these accident types or hazards inflict injury on a worker?" There are 11 basic types of accidents:

- Struck-against
- Struck-by
- Contact-with
- Contacted-by
- Caught-in
- Caught-on
- Caught-between
- Fall-same-level
- Fall-to-below
- Overexertion
- Exposure

Look at each of these basic accident types in more detail. Analyze each job step separately; look for only one kind of hazard or accident at a time.

Why Consider Human Problems in the JSA/JHA Process?

This is why there is a need to consider the following points related to human problems:

- What effects could there be if equipment is used incorrectly?
- Can the worker take shortcuts to avoid difficult, lengthy, or uncomfortable procedures?

Normally, the job steps for a JSA/JHA are listed in a logical sequence. Some workers, however, may wish to change the sequence for one reason or another. For example, one operator may choose to check fluid levels before he/she does a general walk-around. This type of flexibility is good for worker morale and productivity. But, on the other hand, there are times the sequence of the job steps or deviations from the job steps are critical to safe performance of the job. An example of this is that the walk-around inspection must be made and safety deficiencies corrected before the machine is put into service for the day. It would not be safe or proper to do the walk-around inspection after the machine has been put into service.

Why Must Ways to Eliminate or Control Hazards Be Developed?

There are ways to eliminate the hazards by choosing a different process, modifying an existing process, substituting with a less hazardous substance, or improving the environment. In developing solutions, the following steps can be taken:

- Find a new way to do the job.
- Change the physical conditions that create the hazard.
- Change the procedure to eliminate the hazard.
- Reduce the frequency of job performance. Reduce exposure.
- Contain the hazard with an enclosure, a barrier, or guards.
- Use protective devices such as PPE.

Why Change Job Procedures?

If the sequence of job steps or the deviations from established job steps are critical to the safe performance of a job, this should be noted in the JSA/JHA. The next part of the JSA process is to develop the recommended safe job procedures to eliminate or reduce potential accidents or hazards that have been identified. The following four points should also be considered for each hazard identified for the job step:

1. Can a less hazardous way to do the job be found?
2. Can an engineering revision take place to make the job or work area safer?
3. Is there a better way to do the job? This requires determining the work goal and then analyzing various ways to reach the goal to see which way is safest.
4. Are there work-saving tools and equipment available that can make the job safer?

What to consider with this group of questions is how the equipment and work area can be changed, or provided with additional tools or equipment, to make the job safer.

Why Change the Frequency of Performing a Job?

Can the physical conditions that created the hazard be changed? Physical conditions may be tools, materials, and equipment that may not be right for the job. These conditions can be corrected by either engineering revisions or administrative revisions, or a combination of both.

If hazards cannot be engineered out of the job, can the job procedure be changed? Be careful here because changes in job procedures to help eliminate the hazards must be carefully studied. If the job changes are too difficult, long, or uncomfortable, then the employee will take risks or shortcuts to avoid these procedures. Caution must be exercised when changing job procedures to avoid creating additional hazards. Can the necessity of doing the job, or the frequency of performing the job, be reduced? Often, maintenance jobs requiring frequent service or repair of equipment are hazardous. To reduce the necessity of such a repetitive job, ask what can be done to eliminate the cause or condition that makes excessive repair or service necessary.

Why Consider PPE?

Finally, can PPE be used? The use of PPE should always be the last consideration in reducing the hazards of a job. The usefulness of PPE depends entirely on the worker's willingness to use it faithfully. It is always better to control the hazards of a job by administrative or engineering revisions. PPE should only be considered as a temporary solution for protecting a worker from a hazard, or as supplemental protection to other solutions.

During the JSA/JHA process, safety problems are going to surface. Some of these problems will be easily solved with suggestions made to upper management. Administrative revisions are the easiest to make because there is little if any capital outlay. New, better, or additional PPE normally takes minimum expenditure and can be instituted promptly. Work-saving tools and other equipment may take large expenditures and might be phased in over time as tools or equipment is replaced. Engineering revisions may take time to design and install. Changes in physical conditions may have to be engineered into the next upgrade or redesign.

Summary

The steps involved in a JSA/JHA process have been outlined in the previous pages. It should be especially clear that the main point of doing a JSA/JHA is to prevent accidents by anticipating and eliminating hazards. JSA/JHA is a procedure for determining the sequence of basic job steps, identifying potential accidents or hazards, and developing recommended safe job procedures.

JSA/JHA is an accident prevention technique used in many successful safety programs. The JSA/JHA process is not difficult if it is taken with a commonsense approach on a step-by-step basis. The JSA/JHA should be reviewed often and updated with input from both supervisors and workers who do the job every day. The implementation of the JSA/JHA process will mean continuous safety improvements at your workplace with the ultimate goal of zero accidents. JSA takes a little extra effort, but the results are positive and helpful for everybody.

There are many advantages in using JSA/JHA. It provides training to new employees on safety rules and specific instructions on them and how the rules are to be applied to their work. This training is provided before the new employees perform the job task(s). The JSA/JHA also instructs new employees on safe work procedures.

With JSA/JHA, experienced employees can maintain safety awareness behavior and receive clear instructions for job changes or new jobs. Benefits also include updating current safety procedures and instructions for infrequently performed jobs. It is important to involve workers in the JSA/JHA process. Workers are familiar with the jobs and can combine their experience to develop the JSA. This results in a more thorough analysis of the job. A complete JSA/JHA program is a continuing effort to analyze one hazardous job after another until all jobs with sequential steps have a written JSA/JHA. Once established, the standard procedures should be followed by all employees.

Further Readings

Reese, C.D. *Accident/Incident Prevention Techniques (Second Edition)*. Boca Raton, FL: CRC Press, 2012.

Reese, C.D. *Occupational Health and Safety Management (Third Edition)*. Boca Raton, FL: CRC Press, 2016.

Reese, C.D. and J.V. Eidson. *Handbook of OSHA Construction Safety & Health (Second Edition)*. Boca Raton, FL: CRC/Lewis Publishers, 2006.

United States Department of Labor, Mine Safety and Health Administration, *Accident Prevention (Safety Manual No. 4)*. Beckley, WV: US Department of Labor, Revised 1990.

United States Department of Labor, Mine Safety and Health Administration, *Job Safety Analysis: A Practical Approach (Instruction Guide No. 83)*. Beckley, WV: US Department of Labor, 1990.

United States Department of Labor, Mine Safety and Health Administration, *Job Safety Analysis (Safety Manual No. 5)*. Beckley, WV: US Department of Labor, Revised 1990.

United States Department of Labor, Mine Safety and Health Administration. *Safety Observation (MSHA IG 84)*. Beckley, WV: US Department of Labor, Revised 1991.

United States Department of Labor, National Mine Health and Safety Academy. *Accident Prevention Techniques: Job Safety Analysis*. Beckley, WV: US Department of Labor, 1984.

United States Department of Labor, Occupational Safety and Health Administration. *Job Hazard Analysis, (OSHA 3071)*. Washington, DC: US Department of Labor, 1992.

United States Department of Labor, Occupational Safety and Health Administration, Office of Training and Education. *OSHA Voluntary Compliance Outreach Program: Instructors Reference Manual*. Des Plaines, IL: US Department of Labor, 1993.

45

Job Safety Observation

Why is job safety observation (JSO) one of the accident prevention techniques that can be used to assess safe work performance?

There are many categories of accident causes and many terms used to describe these causes. To precisely determine the causes for each category, the terms *person causes* and *environmental causes* are often used. The *actual* and *potential* causes of accidents are generally accepted as the key factors in a successful loss prevention effort. Actual causes—direct and indirect—can only be considered after an accident has occurred. They can be found by asking the question, "What caused the accident?" Potential causes may be avoided before an accident actually occurs by asking the question, "What unsafe conditions (environmental causes) and/or unsafe procedures (person causes) could cause an accident?" Working with actual causes is similar to firefighting, with after-the-fact analysis and hindsight. The process of understanding, determining, and correcting potential causes is comparable to fire prevention or foresight.

All categories of accident causes must be considered and used in any effective loss prevention program. Safety observations and inspections are necessary phases in the overall safety effort. A JSO is the process of watching a person (worker) perform a specific job to detect unsafe behavior or recognize safe job performance. Making a safety inspection is the process of visually examining the work area and work equipment to detect unsafe conditions (environmental causes). Detecting and eliminating potential causes of accidents may best be accomplished when supervisors understand safety observations and when safety inspections become separate phases of the loss prevention work. This chapter deals primarily with making JSOs.

The JSO phase is initiated when a written set of procedures is prepared by management and safety health personnel. The procedures should include a prepared job safety analysis ready for use, step-by-step safe job operating procedures, and the training of all supervisors in observation procedures. Objectives must be established for each step of the JSO program. The establishment of definite goals at all levels of management will give direction to the safety and health effort. This is why management should outline the purpose and types of JSOs, and how to select a job or task for JSO.

Why JSO?

The basic idea of JSO is simple. It is a special effort to see how employees do their jobs. Planned safety observation involves more effort than an occasional or incidental observation of job procedures. This chapter describes various ways of determining unsafe practices and violations of safety rules. Use of a paper form as a record of a JSO allows for the following:

- Determination of safe job procedures
- Determination of the effectiveness of training
- Documentation of other potential hazards
- Prevention prior to an incident or accident
- Identification of potential shortcuts that may result in harm to workers
- Coaching in real time of workers conducting job tasks
- Assessment of both experienced and inexperienced workers
- Immediate feedback to workers
- A permanent record of prevention efforts

This is why JSO as an accident prevention method emphasizes the importance of a proper supervisor–employee relationship. Becoming more interested in the employee through observations will lead to greater cooperation in the safety and health program. The basic tool for making a JSO is job safety/hazard analysis (JSA/JHA). If a JSA/JHA is not used, the supervisor must be completely familiar with the job steps, job hazards, and safe job procedures. The supervisor should observe the employee doing a complete job cycle, paying attention to safe or unsafe procedures and conditions. A planned JSO form should be used.

Why Select a Job or Task for a Planned JSO?

To conduct a planned safety observation, a method must be established to select jobs to observe. The same selection method used in JSA can be used in the job observation selection. The reasons for selection are divided into four areas, listed as follows:

- Select jobs or tasks where accidents have occurred. Jobs may be selected where there have been repeated accidents or lost time due to accidents, or where accidents have required medical treatment. Other selections may be jobs that are not done where the hazards may not be fully known.
- Supervisors may have reasons for observing certain employees. For example, repeat offenses or observed unsafe behavior are examples of an area where a job would be selected because of individual needs. These may be workers who take unnecessary chances, violate safety rules, and develop improper work methods. Workers suspected of being physically or mentally incapable of safe work may also fit into the area of individual needs.
- Many accidents involve experienced employees, and planned safety observations can detect the reasons. Some workers who have been doing a job for many years will often develop shortcuts and effort-saving practices that are hazardous. Inexperienced workers usually follow examples set by supervisors or experienced

workers. When these people work in an unsafe manner, it has an influence on the inexperienced worker.

- Detecting unsafe behavior and practices quickly in the inexperienced employee is important and will allow the supervisor to take corrective measures quickly. This will prevent accidents and will deter employees from developing unsafe work habits. It is much easier and more effective to correct an inexperienced worker when he/she is observed performing unsafely. Further instruction and guidance is usually the required procedure after observations show the need for additional safety training.

Why Preparing for a JSO Is Necessary?

A JSO gives supervisors a positive means of determining the effectiveness of safety instructions. It also serves as a learning tool for the supervisor. The supervisor learns more about each job, each worker, and areas requiring closer supervision. The supervisor will learn to be more perceptive in all areas of responsibility, which will result in the best distribution of supervisory time. The supervisor will learn to devote time and effort where needed. The primary need to accomplish an ongoing observation system and continually improve the safety and health program starts with a written procedure for the JSO. The tailoring of the JSO program to the needs of the operation includes the following:

1. The reasons for the plan (must include orientation of all employees about the purpose of the program)
2. The objectives of the program
3. Training requirements for all levels of supervision
4. Safety and health department's responsibilities
5. Preparation procedures for making a JSO
6. How to make a planned safety observation
7. Recording the observation outcomes
8. Holding a postobservation conference
9. Follow-up procedures (including enforcement policy)

Since supervising workers means observing, the supervisor is the most qualified to make JSOs. He/she knows the workers and the training they have received, and knows the jobs and how these jobs should be done. With proper preparation, the supervisor can conduct a thorough observation. Other management personnel may take part in the planned safety observation, but the supervisor must ultimately be responsible for observation.

In determining why other aspects are chosen for observation, one needs to consider whether (1) a job involves some new procedures because of a recent JSA revision, (2) there has been a change in equipment or machinery, (3) the job poses an exceptional hazard, or (4) it is a job infrequently done but is complex.

Why a Checklist of Activities to Be Observed Is Needed

A checklist or form should be used for recording observations. This form should record what the worker has been observed doing, when the observation was made, and what action or actions were taken or planned as a result. The checklist should include key points of the particular job. If a JSA/JHA is available, it should be used as the basic information and guideline for the observation. The supervisor should watch each step of the job and be alert for compliance with safe job procedures. The supervisor should recognize the need to revise the JSA/JHA if necessary.

A list of possible unsafe actions for the job would provide a guide to use. The supervisor must keep an open mind for possibilities that are not on the list. Most checklists do not cover every conceivable unsafe action. A checklist should also include specific safety behaviors to look for, such as the following:

- How does the worker handle tools and equipment?
- Does the worker show concern for doing a good job?

A checklist can also detect hazards caused by inappropriate clothing and inadequate personal protective equipment. Whether clothing should be considered hazardous may depend on the type of work done, the work area, the weather, or company regulations. A checklist may detect hazards such as the following:

1. Loose or ragged clothing that could be caught in moving machinery
2. Rings, bracelets, necklaces, or wristwatches that could be hazardous when climbing, using certain tools, working around heat or chemicals, or doing electrical work
3. Long hair that could catch in or on moving or rotating machinery or that could become harmful in other ways

Personal protective equipment can also be checked during observations. Items can include the following:

- Proper head protection, eye protection, and foot protection
- When needed, proper gloves, respirators, and suitable ear protection
- The use of protective clothing
- Respirators and other items required for specific areas

Any specific unsafe behavior on the job will correspond to basic types of unsafe procedures. The following unsafe procedures should be included in an observation checklist:

1. Operating or using equipment without authority
2. Failure to secure against unexpected movement
3. Failure to utilize lockout/tagout or blocking
4. Operating or working at an unsafe speed
5. Using unsafe tools and equipment

6. Using tools and equipment unsafely

7. Failure to warn or signal as required

8. Assuming an unsafe position or unsafe posture

9. Removing or making safety devices inoperable

10. Repairing, servicing, or riding equipment in a hazardous manner

11. Failure to wear required personal protective equipment

12. Wearing unsafe personal attire

13. Violation of known safety rules and safe job procedures

14. Indulging in practical jokes, fighting, or sleeping; creating distractions; etc.

Why the Observation?

The employee observed should be aware of the observation from time to time. Inform all employees, as part of their orientation, that JSOs are a phase of the safety and health program. All employees must fully understand the reason for incidental and planned observations. With this in mind, there is another consideration in preparing for the JSO. Should you inform the employee in advance about the observation? Under certain conditions, the employee should have advanced notice.

If trying to find out what an employee knows and does not know about job safety procedures, tell the employee in advance. If any unsafe practices are revealed under this condition, the supervisor can conclude that the employee does not know safety procedures. If the supervisor tells the employee in advance, he/she will learn what is known or not known about job safety procedures.

If the employee is new, or one who has relatively little experience, tell him/her in advance. This is also true for an employee who may have experience but has never been checked using a JSO.

Do not tell the employee in advance about the observation if the objective is to learn how the employee normally does the job. When the supervisor knows that an employee understands how to do a job safely, the supervisor must then find out how the employee works when no one is observing. To determine this, the employee should not be told in advance about the JSO. Most employees work safely under the eyes of the supervisor, especially when they are told of the observation. If the employee does not show unsafe practices, the supervisor can assume that the employee usually works safely. If the same person is later observed doing some part of the job unsafely, the supervisor must believe that work is performed unsafely at other times. The supervisor may know that the employee can perform the job safely from past observation. From observing employees without their knowledge, the supervisor may learn that they are not putting into practice their job knowledge.

The supervisor should never try to observe an employee from a hidden position. The supervisor may learn by the observation, but respect and human relations may be lost. After informing the employee of the observation, the supervisor must stand clear of the employee's work area. The employee must be given plenty of room to work. The supervisor must not create a hazard by standing too close or distracting the employee.

The supervisor should avoid work interruptions unless absolutely necessary. It is better to save all minor corrections until the observation is complete. However, if the employee does something in an unsafe manner, stop him/her immediately. The supervisor should explain what has been done wrong and explain why it is unsafe. Call the employee's attention to the JSA/JHA if training was originally conducted within the framework of a JSA/JHA program. Check to see if the employee fully understands the explanation. This can often be done by having the employee tell and show the safe way to do the job.

Why Conduct a Postobservation?

A brief postobservation discussion should be held with the employee. One of two things should be done. If the job was done as required, the employee should be complimented. Recognition, when deserved and when sincerely given, will reinforce safe behavior. This acts as a positive motivation for the worker. Any unsafe behavior seen during the observation should be discussed. If the employee performed in an unsafe manner, the supervisor should discuss what occurred and get the employee's reasons for doing the job as it was done.

The concern now is not with what has been done but with changing the employee's behavior so the same thing is not done in the future. The supervisor must take required corrective action and should give the reasons behind the measures. The postobservation conference should always end with the feeling that things are right between the supervisor and the employee.

Why Deal with Unsafe Behaviors or Poor Performance?

Dealing with unsafe performance may be helpful in determining why unsafe practices are continuing. A step-by-step procedure to follow in correcting unsafe performance can be developed. The supervisor should record the specific unsafe performance that has been observed and determine the type of accident that could result from this action. Since this action and the possible results have already been discussed with the employee after or during the first observation, the supervisor must decide what has caused this action to continue.

- The first step is to identify the performance problem. In this area, the performance problem may be someone's actual performance and his/her desired performance.

- Is the problem worth solving? Not every performance problem is worth solving. The problem needs to be evaluated realistically. What impact will the performance problem have on doing the job safely? If the problem has or can have no adverse impact, then ignore it.

- In determining the cause, you need to ask about or try to find out the cause of the performance problem. Could this person do it if he/she really had to? If the answer to this question is *no*, then the question needs to be asked, does the employee have the potential to perform this job safely? Both physical and mental capabilities must be assessed in this area. If the answer to this question is *no*, then the job may have to be changed to fit the employee's ability, or if possible, the employee can be placed in a job that fits his/her capabilities. If the employee has the potential to do the job, then the question needs to be asked, has the person ever performed the job? If the answer to the question is *no*, then classroom or on-the-job training will be necessary. If the person has performed the job, refresher training may be all that is necessary.

- If the employee could do the job if he/she really had to, then a nontraining solution will help to identify and solve the performance problem. One way to solve the nontraining problem is to provide feedback. Feedback should come from management being aware of the problem, observing the employee, or simply making the employee aware that he/she is expected to do the job correctly.

- Another way to solve the problem is to eliminate constraints or barriers. Tools, equipment, and time might be in this category since these constraints can take many forms. Ergonomic considerations in job design are an example of a potential constraint. Employees forced to stoop, bend, twist, turn, stretch, lean, reach, or assume unnatural postures to perform routine tasks are subject to discomfort and fatigue. Discomfort and fatigue may distract from immediate hazards, and chronic physical stresses may make employees prime candidates for cumulative trauma disorders such as carpal tunnel syndrome and tendinitis.

- Other constraints or barriers that could adversely impact performance include conflicting demands on the employee's time, authority, and proper tools to perform the required task. Are there incentives at work? Is nonperformance rewarded? This may occur if the job is easier to do unsafely or the employee gets a break if the job is done quickly. Management needs to address this reward problem by eliminating any rewards for poor performance. One way to do this is for management to make sure their only concern is not just that the job is completed but that the employee is not punished if the job takes longer than normal.

- Is good performance punished? One way of punishing good work is by assigning additional work to that "good" employee. Management needs to eliminate "good performance punishment" by eliminating the negative effects and create, or increase the strength of, positive or desirable consequences.

- Is there a positive change? If a positive change occurs, then this change has to be maintained through periodic feedback, praise, and recognition. This must be done on a continuing basis to keep the behavior at a constant level.

- If a positive change does not occur, other actions may need to take place. It has to be determined whether there are meaningful actions for the desired performance. If there are no meaningful actions, then the remedy may be to arrange to take actions to make safety matter to the employee. This may be disciplinary action.

JSO that is implemented properly provides the foundation for addressing unsafe performance, which will equate to continuous improvement with regard to workplace safety and health.

Summary

The primary purpose of a JSO is to help each employee become willing to follow safe job procedures. When finding violations and unsafe practices, the object is not merely to stop the employee from continuing this behavior but to get all employees to follow procedures and to remain productive and cooperative. Safety observation enables the supervisor to evaluate all aspects of an employee's performance on a specific job. An observation that reveals an acceptable performance level provides an opportunity to give praise and reinforce good performance. A safety observation of substandard performance provides an opportunity to take corrective action before unnecessary problems or losses occur.

All companies must establish and enforce rules of conduct and rules of safety for everyone involved in their operation. Since supervisors are responsible for following safety and health rules, a JSO is a valuable method for accomplishing an effective JSO program that can enhance all aspects of the accident/incident program such as JSA, safety inspections, and safety orientation and safety and health training programs. This why JSOs should be an integral part of any occupational safety and health (OSH) program.

Further Readings

Kavianian, H.R. and C.A. Wentz, Jr. *Occupational and Environmental Safety Engineering and Management.* New York: Van Nostrand Reinhold, 1990.

Mager, R.F. *Analyzing Performance Problems.* Belmont, CA: Fearson Publishers, 1970.

Reese, C.D. *Accident/Incident Prevention Techniques (Second Edition).* Boca Raton, FL: CRC Press, 2012.

Reese, C.D. *Occupational Health and Safety Management (Third Edition).* Boca Raton, FL: CRC Press, 2016.

Reese, C.D. and J.V. Eidson. *Handbook of OSHA Construction Safety & Health (Second Edition).* Boca Raton, FL: CRC/Lewis Publishers, 2006.

United States Department of Labor, Mine Safety and Health Administration. *Accident Investigation (Safety Manual No. 10).* Beckley, WV: US Department of Labor, Reprinted 1989.

United States Department of Labor, Mine Safety and Health Administration. *Accident Prevention (Safety Manual No. 4).* Beckley, WV: US Department of Labor, Revised 1990.

United States Department of Labor, Mine Safety and Health Administration. *Job Safety Analysis: A Practical Approach (Instruction Guide No. 83).* Beckley, WV: US Department of Labor, 1990.

United States Department of Labor, Mine Safety and Health Administration. *Job Safety Analysis (Safety Manual No. 5).* Beckley, WV: US Department of Labor, Revised 1990.

United States Department of Labor, Mine Safety and Health Administration, *Safety Observation (MSHA IG 84).* Beckley, WV: US Department of Labor, Revised 1991.

United States Department of Labor, National Mine Health and Safety Academy. *Accident Prevention Techniques: Job Observation.* Beckley, WV: US Department of Labor, 1984.

46

Fleet Safety

When an employer makes a large capital investment in a number of automobiles, trucks, long-haul vehicles, commercial boats, airplanes, off-road construction vehicles, police cruisers, fire trucks, etc., the cost is often formidable and should be taken care of and protected to ensure long-term operation and longevity of the investment. With vehicles that are mobile, continuous maintenance must be performed to insure their safe operation. Also, a process to obtain, hire, and train the best and safest operators (drivers) is part of the process to assure that safe operation is occurring. These two components are critical to protecting these expensive vehicles. Thus, the reason for a preventive maintenance program that cares for vehicles and assures their safe operation is just as critical as the hiring of successful, safe, and responsible drivers and operators.

Any problem that arises with a vehicle in the fleet is usually an expensive undertaking. This is why regular maintenance is invaluable in protecting such from long-term wear and tear. Replacing worn parts will prevent potential failures from causing accidents and incidents, which can be more costly due to damage of the vehicle or injury/death of equipment operators.

Fleet safety is often viewed as operator safety, which is definitely a key component of a company's attempt to protect its large dollar investment in vehicles and mobile equipment. It goes without saying that many of the accidents that occur are a direct result of driver error. But driver error is not the fault of the individual. It is the fault of management's failure to institute a fleet safety program that provides organization, direction, and accountability for the fleet of vehicles that the company owns.

Commitment to a fleet safety program communicates the value that the company places upon their property and employees. The care given to both vehicles/equipment and employees conveys the company's true view of the value of accident prevention.

A fleet safety program should consist of the following:

- Written fleet safety program.
- Vehicle/equipment maintenance procedure.
- Record-keeping process.
- Operator selection process.
- Operator training requirements.
- Operator performance requirements.
- Operators' health and physical well-being.

This is why fleet safety is usually viewed as an accident prevention technique. It is a systematic approach to addressing the safe operation of a fleet of vehicles.

Why Have a Fleet Safety Program?

The company should clearly state their policy regarding fleet safety and delineate what is expected to transpire as a result of its program. The overall intent of this program must be stated clearly. The program should incorporate the many facets of any good accident prevention effort. The fleet safety program should provide the framework for safety management of the company's vehicles/equipment and employees. The company needs to communicate program goals to drivers and supervisory personnel.

There must be a designated person with responsibility for safety and compliance with regulations. This designated person must assume responsibility to comply with existing regulations and implement and enforce the company rules and policies. This person must oversee the qualifying of operators/drivers and the care and safety maintenance of the company's fleet.

It is management's responsibility to recruit and screen new drivers, monitor driver qualifications and safety infractions, and provide training to upgrade drivers' skills and knowledge. The risk involved when drivers/operators are improperly hired or trained is a large liability for a company. These drivers/operators are expected to drive and operate a vehicle in a responsible and safe manner as they drive across public highways, move heavy or awkward loads in a safe manner, drive in precarious off-road situations, or transport hazardous materials or waste. Some more progressive companies are even using driving simulators to see if drivers/operators can react to actual driving situations without having to experience or react to live on-the-job events and scenarios that would be risker and result in more exposure. These simulators can be used for refresher training, postaccident training, or skill determination.

Management should provide a formal mechanism for investigating and reviewing accidents and monitoring maintenance and equipment safety. Management should also implement safe driving incentives and offer recognition to drivers who meet the required standard of performance. The company must constantly monitor the effectiveness of their fleet safety program.

Why Vehicle Maintenance?

The cost of a fleet of vehicles is a staggering investment and a major cash outlay for companies. In order to reap the full benefits of this investment, start with a thorough purchasing process. A company wants quality and dependability for their money. This will entail some research on the company's part to assure that they are getting the most for their money. Once a fleet is in place, then the company will want to get the most mileage out of its purchase. This can only be accomplished by having a preventive maintenance program in place, which includes regularly scheduled maintenance, follow-up to operator complaints, and daily preshift inspections of vehicles and equipment. There should also be a record-keeping system for the maintenance program, which includes the following:

- Operator's inspection record—a checklist of things to be checked daily by operators and any corrections needed to ensure the safety of the vehicle. (This should go to the maintenance shop.)

- Schedule maintenance record—includes maintenance shop records of routine or periodic service for each vehicle.

- Service record—to show all findings and results of the inspections, routine service, and repairs made along with the date of each such maintenance procedure.

- Vehicle history record—a complete history of the vehicle including, but not limited to, any accidents in which it was involved, any catastrophic failure or repairs (i.e., engine change), and when tires were replaced.

Why Is Driver/Operator Selection Important?

The drivers/operators have the most impact on preventing accidents/incidents in operating the vehicles within a fleet. Fleet safety may be viewed as vehicle safety or mechanical safety, but this type of safety depends upon both maintenance and operators. An operator's job must include preoperation inspections and, upon completing his/her use of the vehicle, the reporting of any defects. This should be normal operating procedure for any preventive maintenance program. Nevertheless, what can be expected is dependent upon the quality of the drivers/operators. Operators are the center of the fleet safety program.

Thus, in a fleet safety program, it is important to select the best operators for the job. The operator is vital to the prevention of accidents, incidents, vehicle damage, and injuries. Careful selection of the operator is paramount to an effective fleet safety program. The selection process should involve access to an operator's past employment history, driving record (including accidents), commendations, and awards, as well as previous operator's experience.

This is why as a condition of employment and based upon the criteria in a written job description, all potential operators should be able to pass a physical and mental examination and an alcohol/drug test.

To improve fleet safety, adequately qualified drivers must be recruited and their performance monitored. The great majority of preventable accidents can be shown to be directly related to the performance of the driver. It is therefore extremely productive to any fleet safety program to carefully conduct new driver selection and adequate monitoring procedures for existing drivers.

An established formal procedure for interviewing, testing, and screening applicants needs to be in place. A defined standard of skill and knowledge should be met by successful applicants. Appropriate methods should be in place to check out previous employment histories and references of all potential operators. A check of the prior driving records of the applicants must be performed.

Once an operator is hired, there should be a formal program for monitoring his/her performance, a periodic review of his/her driving record, and a periodic review of his/her health. Operators should be monitored occasionally for drug and alcohol abuse. A means should be in place for identifying deficiencies in an operator's skills and knowledge, and a procedure should be in place for remedial training. It is well worth the effort to establish a procedure for terminating unqualified operators.

Why and What Records Need to Be Maintained?

There are three types of records that need to be maintained as part of a fleet safety program. They are vehicle, operator, and accident records. Vehicle records have been discussed previously. Driver records should be kept, which include hiring, training, and job performance. Hiring records should include the following:

- Application form completed for screening
- Previous employment record
- Character references
- Results of work reference checks
- Record of violations and accidents
- Results of interviews
- Results of tests given
- Physical examination report
- Information on previous driver training
- Driver performance records
- All accident records

The maintenance of records can be used to justify action taken by the company, when inquiries are made by regulatory agencies, and when legal or liability issues arise. This is a good business practice that helps protect the company.

Summary

A preventable accident is one that occurs because the operator fails to act in a reasonably expected manner to prevent it. In judging whether the operator's actions were reasonable, one seeks to determine whether the operator drove defensively and demonstrated an acceptable level of skill and knowledge. The judgment of what is reasonable can be based on a company-adopted definition, thus establishing a goal for its safety management program.

Fleet safety driving performance is dependent on management's commitment to the implementation of a formal fleet safety program.

This is why the concept of a preventable accident is a fleet safety management tool, which achieves the following goals:

1. It helps to establish a safe driving standard for the driver.
2. It provides a criterion for evaluating individual drivers.
3. It provides an objective for accident investigations and evaluations.
4. It provides a means for evaluating the safety performance of individual drivers and the fleet as a whole.

5. It provides a means for monitoring the effectiveness of fleet safety programs.

6. It assists in dealing with driver safety infractions.

7. It assists in the implementation of safe driving recognition programs.

This is why fleet safety driving performance is dependent on management's commitment to the implementation of a formal fleet safety program. An effective safety program will interact with most aspects of fleet operations and challenge the skills and knowledge of its supervisors and drivers.

Further Readings

National Safety Council. *Motor Fleet Safety Manual (Third Edition)*. Itasca, NY: NSC, 1986.

Reese, C.D. *Accident/Incident Prevention Techniques (Second Edition)*. Boca Raton, FL: CRC Press, 2012.

Reese, C.D. *Occupational Health and Safety Management (Third Edition)*. Boca Raton, FL: CRC Press, 2016.

Reese, C.D. and J.V. Eidson. *Handbook of OSHA Construction Safety & Health (Second Edition)*. Boca Raton, FL: CRC/Lewis Publishers, 2006.

47

Preventive Maintenance Programs

Why is a preventive maintenance program (PMP) necessary? The simplest answer is to prolong the life of vehicles, machinery, and equipment while maintaining their safe operation during that period.

A PMP depends heavily on an inspection form or checklist to assure that a vehicle, machinery, or equipment inspection procedure has been fully accomplished and its completion documented. It has long been noted that a PMP has benefits that extend beyond caring for equipment, vehicles, and machinery. Of course, equipment is expensive and, if cared for properly and regularly, will last a lot longer, cost less to operate, operate more efficiently, and have fewer catastrophic failures.

Remember, properly maintained equipment is also safer, and there is a decreased risk of accidents occurring. The degree of pride in having safe operating equipment will transfer to the workers in the form of better morale and respect for the equipment. Well-maintained equipment sends a strong message regarding safe operation of equipment.

Other factors that provide legitimate reasons to have a PMP are as follows:

- Reduced cost by not having to bear the cost of untimely replacement.
- Maintenance records support and document warrantee issues.
- Provide records if liability issues arise.
- Provide due diligence to demonstrate Occupational Safety and Health Administration (OSHA) compliance.
- PMP records become a component of evidence in case an unforeseen event transpires.
- Creates confidence that any vehicle, machinery, or equipment can be used safely in performing work-related tasks.
- PMPs indicate that the manufacturer's maintenance recommendations and scheduled maintenance have been completed.

If operators are allowed to use equipment, machinery, or vehicles that are unsafe or in poor operating condition, a negative message is sent that says, "I don't value my equipment, machinery, or vehicles, and I don't value my workforce either." A structured PMP will definitely foster a much more positive approach regarding property and the workforce. Reasons for a PMP are as follows:

1. Improve operating efficiency of equipment, machinery, or vehicles
2. Improve attitudes toward safety by maintaining good/safe operating equipment
3. Foster involvement of not only maintenance personnel but also supervisors and operators, which forces everyone to have a degree of ownership
4. Decrease risk for incidents or mishaps

When a problem with any vehicle, machinery, or equipment exists or arises, it should be addressed immediately to preclude further damage or exacerbation of the problem. This is why:

- The first aspect of a PMP is to have a schedule for regular maintenance of all your equipment.
- Second, it motivates supervisors to make certain all operators are conducting daily inspections.
- Third, it assures the company that all defects are reported immediately.
- Last, it documents that repairs are made prior to operating vehicles or equipment. If this is impossible, the equipment should be tagged and removed from service.

Why Specific Components Are Needed for an Effective PMP?

The following are invaluable in order to have an effectively functioning PMP:

1. A maintenance department, which carries out a regular and preventive maintenance schedule
2. Supervisors and operators who are accountable and responsible
3. A preshift checklist for each type of equipment, machinery, or vehicle
4. An effective response system when defects or hazards are discovered
5. A commitment by management that your PMP is important and will be achieved

PMPs are another example of an accident prevention technique. Of course, this technique would apply only if the company has equipment, machinery, or vehicles. A PMP also can be an integral part of a fleet safety program.

Why Preventive Maintenance?

Preventive maintenance's ultimate purpose is to prevent accidents caused by vehicle deficiencies, machinery defects, or equipment hazards. Worn, failed, or incorrectly adjusted components can cause or contribute to accidents. Preventive maintenance and periodic inspection procedures help to prevent failures from occurring while the vehicle/machine/equipment is being operated. Set preventive procedures also reduce reliance on the operator, who may have limited skill and knowledge for detecting these deficiencies.

Why Management's Role?

If management experiences large numbers of repairs to the vehicles, machinery, and equipment, this should be viewed as an indication of inadequate maintenance and inspection procedures and an indication of potentially unsafe vehicles, machinery, or equipment, which could facilitate or cause accidents. The scheduling of periodic inspections and maintenance activities should be part of management's safety and health program. Management should maintain a record on each vehicle regarding repairs, inspections, and maintenance.

When hazards or defects are found, there must be guidelines or rules for placing vehicles, equipment, or machinery out of service until problems are corrected. Operators must be sufficiently trained and knowledgeable enough to detect maintenance and repair needs, and to report them to the appropriate authority.

Is the emphasis of your preventive maintenance and inspection program based upon the importance of recognizing the components that directly affect control, such as braking systems, steering systems, couplers, drives, kill switches, electrical controls, tires and wheels, and suspension systems? Each of these could affect safe operation. You should also have a program that can determine when vehicle, machinery, and equipment component wear is at a point where replacement is necessary.

This is why the PMP must be enforced by management. It should also be evaluated continuously to gauge its effectiveness in preventing accidents. Your fleet of vehicles, equipment, and machinery should be capable of passing the minimum industry inspection standards.

Why a Formalized PMP?

A preventive maintenance and inspection program should recognize wear of consumable components, which must be periodically replaced or serviced. It should take into account indicators of deterioration that can be monitored at the operator inspection level. The operator should be trained in troubleshooting. Special attention must be paid to the condition of components, which cannot be easily observed by the operator. Maintenance supervisors and mechanics should inspect those components where problems can occur but are not easily discernible.

Why the Operator Should Conduct Inspections?

To ensure that vehicles are in safe operating condition while being driven or operated, the operator should check the whole vehicle carefully, preoperation and postoperation. These inspections reports are an important component of a PMP. If something seems to be wrong with the vehicle, equipment, or machinery, stop and check it out. Do not continue with the operation until the operator is satisfied that it is safe to do so.

The operator should be the one held ultimately responsible to make sure that the vehicle, equipment, or machinery is in a safe operating condition. Appropriate inspection procedures and reports assist in ensuring this. The operator is also in a position to detect deficiencies and refer them to maintenance for repairs. The operator should not operate a faulty vehicle, equipment, or machinery. Federal and state laws require that the operator should not drive a vehicle unless fully satisfied that it is in a safe operating condition.

Why Maintenance?

Maintenance crews should use established company and industry guidelines to determine whether vehicles, equipment, and machinery should be returned to service. They need to determine the cause of damage or deterioration. Such analysis may help identify improper use of machinery, equipment, or vehicles, or faulty maintenance procedures, which should be corrected. All maintenance or repairs for each vehicle should be documented on a separate maintenance reporting form.

Never allow any vehicle, machinery, or equipment that is not operating safely to leave the maintenance facility. Unsafe vehicles, machinery, or equipment should be locked out and tagged out to prevent their use.

Why Management's Responsibility?

Management's responsibility is to establish inspection and reporting procedures for drivers. It must assure compliance with state, federal, and industry rules. Its job is to select operators and provide them with adequate training for inspecting critical components and in order to determine whether their vehicle, equipment, or machinery is safe to be put into service. Management must enforce both operator inspection and maintenance policies. It must see to it that maintenance personnel are responsive to reported deficiencies, and it must establish company standards for placing vehicles, equipment, or machinery out of service. It must assure that preventive maintenance procedures adequately detect and repair worn or defective components.

Management must make sure operators and maintenance personnel are trained and knowledgeable to make a determination during inspections as to whether or not a vehicle, machinery, or equipment is safe or should be taken out of service. Neither set of employees should perceive that management wants them to circumvent good safety practices.

Summary

A PMP can be developed for any industrial function that has a maintenance component. The same principles used for the operator and operation can be used for other mobile equipment, machinery, or process maintenance. It matters little the type of maintenance. It is more important that the workers, operators, supervisor, and management realize that the care taken in prevention will pay dividends in cost savings, lower accident cost, less equipment damage, and better overall working conditions and employee attitudes. Care of physical assets (physical plant, vehicles, machinery, etc.) is critical to the bottom line. The bonus is that it will reduce one of the factors that can contribute to the occupational injury and illness rate.

Further Readings

Reese, C.D. *Accident/Incident Prevention Techniques (Second Edition).* Boca Raton, FL: CRC Press, 2012.

Reese, C.D. *Occupational Health and Safety Management (Third Edition).* Boca Raton, FL: CRC Press, 2016.

Reese, C.D. and J.V. Edison. *Handbook of OSHA Construction Safety & Health (Second Edition).* Boca Raton, FL: CRC/Lewis Publishers, 2006.

United States Department of Labor, Occupational Safety and Health Administration, Office of Training and Education. *OSHA Voluntary Compliance Outreach Program: Instructors Reference Manual.* Des Plaines, IL: Department of Labor, 1993.

Section J

Extraneous Hazards

At times, the workplace is impacted by forces, activities, or new/arising events or topics that are extraneous to the initial origins or purpose of the company but become an issue that cannot be overlooked in the functioning of the company. The chapters in this section discuss how these outside hazards affect the overall safety and health of the workplace and workforce.

The contents of this section are as follows:

Chapter 48—Emergency Planning

Chapter 49—Workplace Security and Violence

Chapter 50—External Force: Terrorism

Chapter 51—Off-the-Job Safety

48

Emergency Planning

A workplace emergency is an unforeseen situation that threatens employees, customers, or the public; disrupt or shuts down an operation; or causes physical or environmental damage. Emergencies may be natural or man-made.

Each and every workplace or business has the potential to experience an emergency situation or event. The failure of a company or business to plan for an emergency is actually a plan to fail, when an emergency transpires. This is why an ineffective response to an emergency could result in adverse outcomes. The questions that arise are as follows:

- Why was the emergency not planned for?
- Why was there an inadequate response?
- Why was there not a plan for the specific emergency?
- Why were employees not trained to respond to the emergency?
- Why had the employees not been drilled or practiced regarding response to emergencies?
- Why were there no signals or alarms?
- Why were evacuation routes and exits not marked?
- Why was medical care not readily available?
- Why did emergency response personnel or rescuers fail to respond properly?

There is no way that an industry or business can account and plan for the all the unforeseen emergency events. The failure to at least assess or evaluate the potential for emergencies or at the very least address the types of emergencies that could occur at a specific jobsite or work site could increase the amount of liability faced by the company or business.

It is extremely doubtful that all industry sectors or businesses do not have to plan for an emergency. Where employees exist, there is always potential for human emergencies such as injury, illness, death, violence, and medical emergencies. The weather and environment can also create natural emergencies such as storms, floods, earthquakes, and tornados. The man-made emergencies can not be overlooked either, such as fire, explosion, or chemical spills. The key is to be the best prepared as possible to react to any emergency.

In recent years, emergency response has become much more complicated due to more than fire, accidents, weather, etc. With the advent of terrorism, all types of issues have arisen, such as bombing, sabotage, employee security, structure security, domestic disputes, etc. The potential outcomes of not planning for emergencies are destruction, deaths, injuries, illnesses, damage, property loss, pollution, and catastrophes. This is just another reason why emergency preparedness requires planning.

When looking at what hazards exist in a workplace, it is imperative that a worst-case scenario approach be employed. It is virtually impossible to address each possible hazard, but each industry has areas where it is most vulnerable or most at risk of an unplanned emergency event. A risk assessment of some type is the approach that will allow for the prioritization of the potential risk. From a risk assessment, an action plan can be developed to address the hazards that have been identified. Some of the common hazards that might be identified as having an impact on the workplace and can be causal factors that may facilitate emergencies are as follows:

Potential Causes of Emergencies

- Accidental release of toxic substances
- Accidents (injuries, illnesses, and deaths)
- Airborne chemical or biological releases
- Avalanche
- Bomb threat
- Catastrophic failure
- Chemical spill
- Civil strife
- Collapses
- Deliberate release of hazardous biological agents or toxic substances
- Disease outbreaks (epidemic)
- Earthquakes
- Explosion
- Exposure to ionizing radiation
- Fire
- Flooding
- Foreign travel
- Hurricanes
- Labor strike
- Landslide
- Live conductors
- Loss of communications
- Major structural failure
- Mechanical failure
- Medical events
- Natural gas leak
- Power outage
- Release of radioactive materials
- Sabotage
- Security issues
- Spills of flammable liquids
- Suspicious letter or package
- Tornados
- Transportation incidents (truck, rail, or air)
- Tsunami
- Typhoon
- War or conflict
- Water leak
- Windstorms
- Winter storms
- Workplace violence

Why plan for emergencies? First and foremost, it is a moral obligation. Secondly, it is a legal obligation. Thirdly, government, state, and local regulations require such planning.

The best way is to prepare to respond to an emergency before it happens. Few people can think clearly and logically in a crisis, so it is important to do so in advance, when you have time to be thorough. This is another reason why an emergency action plan (EAP) should be an integral part of emergency planning as an element of good practice for proper safety and health management.

Why EAPs?

The question as to why an employer needs to develop a plan of action to address emergencies is not complex. *Why* begins with preventing fatalities, injuries, and illnesses while reducing damage to buildings, stock, and equipment, the result of which is the acceleration of the resumption of normal operations. It also facilitates a company's ability to recover from financial losses, regulatory fines, and loss of market share. Emergency planning can reduce exposure to civil and criminal liability in the event of an incident. It will reduce insurance premiums while enhancing the company's image and credibility with employees, customers, suppliers, and the community.

An EAP covers designated actions employers and employees must take to ensure employee safety from fire or other potential emergencies. Not all employers are required by the Occupational Safety and Health Administration to establish an EAP, but from a philosophical basis for an effective safety and health initiative, an EAP is a responsible approach for all employers.

Not every employer is required to have an EAP. The occupational safety and health standards that require such plans include the following:

- Process Safety Management of Highly Hazardous Chemicals, 1910.119
- Fixed Extinguishing Systems, General, 1910.160
- Fire Detection Systems, 1910.164
- Grain Handling, 1910.272
- Ethylene Oxide, 1910.1047
- Methylenedianiline, 1910.1050
- 1,3 Butadiene, 1910.1051

If the employer has 10 or fewer employees, then he/she can communicate the plan orally instead of the required written plan that is to be maintained in the workplace and available to all workers. When required, employers must develop EAPs.

When developing an EAP, it's a good idea to look at a wide variety of potential emergencies that could occur in the workplace. It should be tailored to the specific work site and include information about all potential sources of emergencies.

Prior to developing a written EAP, employers will need to assess the workplace to determine the types of emergencies that have the realistic potential to occur with his/her work environment. It is definitely impossible to plan for every conceivable emergency, but planning for the most likely risk will help plan for the unforeseen.

Developing an EAP means the employer should do a hazard assessment to determine what, if any, physical or chemical hazards exist in the workplace and could cause an emergency. If an employer has more than one work site, each site should have an EAP tailored for that site.

Once the EAP has been reviewed with employees and everyone has had the proper training, it is a good idea to hold practice drills as often as necessary to keep employees prepared. Include outside resources such as fire and police departments when possible. After each drill, gather management and employees to evaluate the effectiveness of the drill. Identify the strengths and weaknesses of your plan, and work to improve it. Review your plan with all employees, and consider requiring annual training in the plan. Also offer training when you do the following:

- Develop your initial plan.
- Hire new employees.
- Introduce new equipment, materials, or processes into the workplace that affect evacuation routes.
- Change the layout or design of the facility.
- Revise or update your emergency procedures.

Why Are Certain Elements Required or Recommended?

First and foremost, in an emergency, there should be a mechanism of reporting its occurrence to proper authorities and alerting workers by some recognized alarm system that is both audible and visual (for those hearing impaired). All aspects of the EAP should be conveyed to all workers as well as drilled so that they react as espoused in the plan. The reaction to an emergency, if practiced, becomes innate. These drills will expedite a safe and speedy evacuation and allow for accountability of evacuees. This is the reason to practice emergency response and evacuation.

Emergency routes and exits should have auxiliary lighting be legibly marked, and be free from obstructions. Trained evacuation marshals should oversee evacuations.

Any employee who is expected to respond to an emergency or perform duties expected of workers who are designated to evacuate must be trained and practiced in order to fulfill their duties.

Knowing who is responsible in case of an emergency will afford a trained leader to manage the emergency effectively. The coordinator should be responsible for the following:

- Assessing the situation to determine whether an emergency is present that requires activation of your emergency procedures.
- Supervising all efforts in the area, including evacuating personnel.
- Coordinating outside emergency services, such as medical aid and local fire departments, and ensuring that they are available and notified when necessary.
- Directing the shutdown of work operations when required.
- A preferred method for reporting fires and other emergencies.

- An evacuation policy and procedure, including type of evacuation and exit route assignment.
- Describing the routes for workers to use and procedures to follow; emergency escape procedures and route assignments, such as floor plans, workplace maps, and safe or refuge areas.
- Procedures to account for all employees after evacuation.
- Including procedures for evacuating disabled employees.
- Names, titles, departments, and telephone numbers of individuals both within and outside your company to contact for additional information or explanation of duties and responsibilities under the emergency plan.
- Procedures for employees who remain to perform or shut down critical plant operations, operate fire extinguishers, or perform other essential services that cannot be shut down for every emergency alarm before evacuating.
- Including preferred means of alerting employees to an emergency.
- Providing for an employee alarm system throughout the workplace.
- Requiring an alarm system that includes voice communication or sound signals such as bells, whistles, horns, and flashing strobe lights; using tactile devices to alert employees who would not otherwise be able to recognize an audible or visual alarm.
- Making the evacuation signal known to employees.
- Ensuring emergency training.
- Ensuring that the EAP remains available for employee review.
- Requiring employer review of the plan with new employees and with all employees whenever the plan is changed.
- In addition, employers must designate and train employees to assist in safe and orderly evacuation of other employees. Employers must also review the EAP with each employee when the following occur:
  - The plan is developed or an employee is assigned initially to a job.
  - The employee's responsibilities under the plan change.
  - The plan is changed. Keep it updated.
- Rescue and medical duties for any workers designated to perform them.
- Designating and training employees to assist in a safe and orderly evacuation of other employees, including those who have handicaps.
- The site of an alternative communications center to be used in the event of a fire or explosion.
- A secure on-site or off-site location to store originals or duplicate copies of accounting records, legal documents, your employees' emergency contact lists, and other essential records.
- Providing an updated list of key personnel, such as the manager or physician, in order of priority, to notify in the event of an emergency during off-duty hours.

The EAP needs to be reviewed and used to train each employee when the plan is developed and when an employee is assigned initially to a job, the plan is changed, or employees have specific responsibilities under the plan.

Why Preplan?

Since emergencies will occur, preplanning is necessary to prevent a possible disaster. An urgent need for rapid decisions, short reaction time, and lack of resources (emergency equipment) and trained personnel can lead to chaos during an emergency. Time and circumstances in an emergency may mean that normal channels of authority and communication cannot be relied upon to function as normally intended. The stress of the situation can lead to poor judgment, resulting in severe losses. The preplan may be a rather extensive document with instructions, responsibilities, special procedures, unique equipment needs, designated chain of command, prearrangements with emergency response personnel and their role in an emergency event, first-aid training, and list of adequate medical supplies.

The plan should be a living document that has been preflighted, practiced, and drilled and has been changed based upon these activities. It should be thoroughly reviewed after an emergency and any deficiencies corrected.

Summary

The occurrence of emergencies is always a possibility. Reacting to an emergency after the fact can be disastrous and detrimental to any employer's business. The preplanning can actually save a business by decreasing liability and preventing the harm that can occur to the workforce and the workplace itself. Employers should take emergency planning as a serious undertaking and a good business practice.

Why plan? In the event of a major emergency, it is crucial that everyone understands what their role is and what they need to do. This will help do the following:

- Minimize the effects of an emergency as far as possible
- Prevent the escalation or spread of damage
- Contain the immediate effects
- Preserve essential services
- Minimize liability and damages
- Protect the population and the environment
- Restore normality as quickly as possible

A definite plan to deal with major emergencies is an important element of an occupational safety and health (OSH) program. Besides the major benefit of providing guidance during an emergency, developing the plan has other advantages. It may identify unrecognized hazardous conditions that would aggravate an emergency situation that can be worked on or eliminated. The planning process may bring to light deficiencies, such as the lack of resources (equipment, trained personnel, supplies), or items that can be rectified before an emergency occurs. In addition, an emergency plan promotes safety awareness and shows the company's commitment to the safety of workers. The lack of an emergency plan could lead to severe losses such as multiple casualties and possible financial collapse of the company.

Further Readings

Reese, C.D. *Office Building Safety and Health*. Boca Raton, FL: CRC Press, 2004.

Reese, C.D. *Handbook of Safety and Health for the Service Industry: Volume 1, Industrial Safety and Health for Goods and Materials Services*. Boca Raton, FL: CRC Press/Taylor & Francis Group, 2009.

Reese, C.D. *Handbook of Safety and Health for the Service Industry: Volume 4, Industrial Safety and Health for People-Oriented Services*. Boca Raton, FL: CRC Press/Taylor & Francis Group, 2009.

Reese, C.D. *Accident/Incident Prevention Techniques (Second Edition)*. Boca Raton, FL: CRC Press/Taylor & Francis Group, 2012.

Reese, C.D. *Occupational Health and Safety: A Practical Approach (Third Edition)*. Boca Raton, FL: CRC Press/Taylor & Francis Group, 2016.

United States Department of Health and Human Services, Centers for Disease Control and Prevention, *Occupational Health and Safety Manual*. Atlanta, GA: US Department of Health and Human Services, 2002.

United States Department of Labor: Occupational Safety and Health Administration, *How to Plan for Workplace Emergencies and Evacuations (OSHA 3088)*. Washington, DC: US Department of Labor, 2001.

United States Department of Labor, *Occupational Safety and Health Standards for General Industry (29 CFR 1910)*. Washington, DC: US Department of Labor, 2003.

49

Workplace Security and Violence

In recent years, it has become apparent that workplace security and violence is a major emphasis of occupational safety and health (OSH). Workers should feel that they can come to work and work at their jobs without the threat that they may come to harm in some way from violence during their work shift. With this realization, employers have needed to take steps to try to provide security for their workforce.

Each worker can face violence due to the type of work he/she performs, the type of business, the product produced, or the client/customer of the service. At times, this violence can come from within the workplace from mentally ill workers or disgruntled employees. The outside sources can be clients who have issues with employees such as spouses, a criminal, individuals with personal problems, or groups such as terrorists who have as their purpose violence. This makes the case for violence prevention. But primarily, workers need to be assured that, to best of the employer's ability, he/she is working to make them feel safe while working.

Thus, with the escalation of workplace violence in the past two decades, violence in the workplace has reared its ugly head as a workplace issue, with homicide being the third leading cause of occupational death among all workers in the United States from 1980 to 1988 and the leading cause of fatal occupational injuries among women from 1980 to 1985. Higher rates of occupational homicides were found in the retail and service industries, especially among sales workers. This increased risk may be explained by contact with the public and the handling of money.

Research into the causes of the increasing incidence of death and serious injury to health care workers has led to the theory that exposure to the public may be an important risk. The risk is increased particularly in emotionally charged situations with mentally disturbed persons or when workers appear to be unprotected.

This is why it is the employer's responsibility to provide a workplace free from hazards that could cause death or serious physical harm, and this includes workplace violence. Thus, the employer of today must take into consideration the security of his/her workplace to assure that employees can perform their work without the interference of outside sources of danger.

Workplace Violence Statistics

In 1995, the Workplace Violence Institute reported that the cost of workplace violence was $35.5 billion. In 2005, 355,000 (5%) of workplace violence incidents occurred in the 7.1 million private industry businesses. The fatalities in the workplace in 2005 were 5,702, with 564 of these being homicides. Homicide is the second leading cause of workplace deaths or one in every six fatal occupational injuries. Eight percent were caused by firearms, and 20% were from bombings, stabbings, or beatings.

The US Department of Justice indicates that between 1992 and 1996, 2 million individuals were victims of violent crime or threatened with violent criminal acts in the workplace. Simple assault accounted for 1.5 million of these incidents.

In many cases (37%), the victims of workplace violence knew their offenders, but only 1% were victimized by a current or former spouse, boyfriend, or girlfriend. Among these, a spouse was the perpetrator of the crime 21% of the time when a woman was the victim and 2% of the time when the victim was male.

Of those committing workplace violence, 67% were males and 33% females. The victims were injured only 12% of the time. Of the injured victims, only one-half required medical attention.

The estimated annual victimizations for the years 1993 through 1999 for workplace crimes counted by the Bureau of Justice Statistics' National Crime Victimization Survey and the Bureau of Labor Statistics indicated the following:

- All violent crimes = 1.75 million
- Simple assaults = 1.3 million (74%)
- Aggravated assaults = 325,000 (19%)
- Robberies = 70,100 (4%)
- Rapes and sexual assaults = 36,500 (2%)
- Homicides = 900 (.05%)

The rate of violence per 1,000 workers during the 5-year period for selected occupations was as follows:

- Law enforcement officer = 306
- Prison or jail correction officer = 218
- Taxi driver = 184
- Private security guard = 117
- Bartender = 91
- Mental health professional = 80
- Gas station attendant = 79
- Convenience or liquor store clerk = 68
- Mental health custodial worker = 63
- Junior high or middle school teacher = 57
- Bus driver = 45
- Special education teacher = 41
- High school teacher = 29
- Elementary school teacher = 16
- College or university teacher = 3

Fewer than half of all the nonfatal workplace crimes are reported to the police.

Risk Factors

Some of the common risk factors for workers who could be affected by workplace violence are as follows:

- Contact with the public
- Exchange of money
- Delivery of passengers, goods, or services
- Having a mobile workplace such as a taxi or police cruiser
- Working with unstable or volatile persons in health care, social service, or criminal justice settings
- Working alone or in small numbers
- Working late at night or during early morning hours
- Working in high-crime areas
- Guarding valuable property or possessions
- Working in community-based settings

Risk factors may be viewed from the standpoint of (1) the environment, (2) administrative controls, or (3) behavior strategies.

Prevention Strategies and Security

Usually, there are three main areas that must be considered when looking at attempts to provide security and safety for the workforce due to violent occurrences within and without the workplace. These strategies are a good starting point.

Environmental Design

Commonly implemented cash-handling policies in retail settings include procedures such as using locked drop safes, carrying small amounts of cash, and posting signs and printing notices that limited cash is available. It may also be useful to explore the feasibility of cashless transactions in taxicabs and retail settings through the use of debit or credit cards, especially late at night. These approaches can be used in any setting where cash is currently exchanged between workers and customers.

Physical separation of workers from customers, clients, and the public through the use of bullet-resistant barriers or enclosures has been proposed for retail settings, such as gas stations and convenience stores, hospital emergency departments, and social service agency claims areas. The height and depth of the counters (with or without bullet-resistant barriers) are also important considerations in protecting workers, since they introduce physical distance between workers and potential attackers. Nonetheless, consideration must be given to the continued ease of conducting business: a safety device that increases frustration for workers, customers, clients, or patients may be self-defeating. Visibility and lighting are also important environmental design considerations. Making high-risk areas visible to more people and installing good external lighting should decrease the risk of workplace assaults.

Access to and egress from the workplace are also important areas to assess. The number of entrances and exits, the ease with which nonemployees can gain access to work areas because doors are unlocked, and the number of areas where potential attackers can hide are issues that should be addressed. These issues have implications for the design of buildings and parking areas, landscaping, and the placement of garbage areas, outdoor refrigeration areas, and other storage facilities that workers must use during a work shift.

Numerous security devices may reduce the risk of assaults against workers and facilitate the identification and apprehension of perpetrators. These include closed-circuit cameras, alarms, two-way mirrors, card-key access systems, panic-bar doors locked from the outside only, and trouble lights or geographic locating devices in taxicabs and other mobile workplaces. Personal protective equipment such as body armor has been used effectively by public safety personnel to mitigate the effects of workplace violence. For example, the lives of more than 1,800 police officers have been saved by Kevlar vests.

Administrative Controls

Staffing plans and work practices (such as escorting customers and visitors and prohibiting unsupervised movement within and between work areas) are issues that need to be addressed regarding security. Increasing the number of staff on duty may also be appropriate in any number of service and retail settings. The use of security guards or receptionists to screen persons entering the workplace and controlling access to actual work areas has also been suggested by security experts.

Work practices and staffing patterns during the opening and closing of establishments and during money drops and pickups should be carefully reviewed for the increased risk of assault they pose to workers. These practices include having workers take out garbage, dispose of grease, store food or other items in external storage areas, and transport or store money.

Policies and procedures for assessing and reporting threats allow employers to track and assess threats and violent incidents in the workplace. Such policies clearly indicate a zero tolerance of workplace violence and provide mechanisms by which incidents can be reported and handled. In addition, such information allows employers to assess whether prevention strategies are appropriate and effective. These policies should also include guidance on recognizing the potential for violence, methods for defusing or deescalating potentially violent situations, and instruction about the use of security devices and protective equipment. Procedures for obtaining medical care and psychological support following violent incidents should also be addressed. Training and education efforts are clearly needed to accompany such policies.

Behavioral Strategies

Training employees in nonviolent response and conflict resolution has been suggested to reduce the risk that volatile situations will escalate to physical violence. Also critical is training that addresses hazards associated with specific tasks or work sites and relevant prevention strategies. Training should not be regarded as the sole prevention strategy but as a component in a comprehensive approach to reducing workplace violence. To increase vigilance and compliance with stated violence prevention policies, training should emphasize the appropriate use and maintenance of protective equipment, adherence to administrative controls, and increased knowledge and awareness of the risk of workplace violence.

Perpetrator and Victim Profile

Only a small percentage of violence is perpetrated by the mentally ill, gang members, distraught relatives, drug users, social deviants, or threatened individuals, who are often aggressive or violent. A history of violent behavior is one of the best indicators of future violence by an individual. This information, however, may not be available, especially for new workers, patients, or clients. Even if this information were available, workers not directly involved with these individuals would not have access to it.

Workers who make home visits or do community work cannot control the conditions in the community and have little control over the individuals they may encounter in their work. The victims of assault are often untrained and unprepared to evaluate escalating behavior or to know and practice methods of defusing hostility or protecting themselves from violence. Training, when provided, is often not required as part of the job and may be offered infrequently. However, using training as the sole safety program element creates an impossible burden on the employee of safety and security for himself/herself, coworkers, or other clients. Personal protective measures may be needed, and communication devices are often lacking.

Cost of Violence Is a Reason to Address It

Little has been done to study the cost to employers and employees of work-related injuries and illnesses, including assaults. A few studies have shown an increase in assaults over the past two decades. In one reported situation of 121 workers sustaining 134 injuries, 43% involved lost time from work, with 13% of those injured missing more than 21 days from work. In this same investigation, an estimate of the costs of assault was that the 134 injuries from patient violence cost $766,000 and resulted in 4,291 days lost and 1,445 days of restricted duty.

Additional costs may result from security or response team time, employee assistance program or other counseling services, facility repairs, training and support services for the unit involved, modified duty, and reduction of effectiveness of work productivity in all staff due to a heightened awareness of the potential for violence. The cost of not developing and providing security at an employer's workplace could be disastrous to the business. Hence, it is imperative that a part of the OSH effort be directed toward security and the prevention of workplace violence.

Prevention Efforts

Although it is difficult to pinpoint specific causes and solutions for the increase in violence in the workplace, recognition of the problem is a beginning. Some solutions to the overall reduction of violence in this country may be found in actions such as eliminating violence in television programs, implementing effective programs of gun control, and reducing drug and alcohol abuse. All companies should investigate programs recently instituted by several convenience store chains or robbery deterrence strategies such as increased lighting, closed-circuit TV monitors, and visible money-handling locations. If sales are involved, consider limiting access and egress and providing security staff. An employer might want to construct a response plan. Although it may not help to prevent incidents, a response plan should be incorporated into an overall plan of prevention. Training employees in management of assaultive behavior (MAB) or professional assault response has been shown to reduce the incidence of assaults. Administrative controls and mechanical devices are being recommended and gradually implemented.

Some safety measures may seem expensive or difficult to implement but are needed to adequately protect the health and well-being of workers. It is also important to recognize that the belief that certain risks are part of the job contributes to the continuation of violence and possibly the shortage of trained workers.

These guidelines, while not exhaustive, include philosophical approaches as well as practical methods to prevent and control assaults. The potential for violence may always exist for workers; the cooperation and commitment of employers are necessary. However, it is a must to translate these guidelines into an effective program for the occupational safety and health of the workforce.

Why Program Development and Essential Elements?

In order to be consistent with the earlier suggestions, a safety and health program needs to be developed, and the four critical elements in a safety and health program need to be addressed to demonstrate how to put safety and health program development to use related to violence and security as the subject. This is why employers must develop a written program to deter violence and provide security for its workforce. The program should include the following:

Management commitment and employee involvement
   Commitment by top management
   Employee involvement
Hazard identification and analysis
   Record review
   Identification of security hazards
   Workplace violence and security assessment
Hazard prevention and control
   General building, workstation, and area designs
   Maintenance
   Engineering control
   Administrative controls and work practices
Training and education
   Participants in the training program
   Job-specific training

> *Initial training program*
> *Training for supervisors and managers, maintenance and security personnel*

Medical management

Record keeping

Evaluation of the program

Types of Workplace Violence and Events

When one examines the circumstances associated with workplace violence, events can be divided into three major types. However, it is important to keep in mind that a particular occupation or workplace may be subject to more than one type. In all three types of workplace violence events, a human being, or hazardous agent, commits the assault.

In type I, the agent has no legitimate business relationship to the workplace and usually enters the affected workplace to commit a robbery or other criminal acts. In type II, the agent is either the recipient or the object of a service provided by the affected workplace or the victim, e.g., the assailant is a current or former client, patient, customer, passenger, criminal suspect, inmate, or prisoner. In type III, the agent has some employment-related involvement with the affected workplace. Usually, this involves an assault by a current or former employee, supervisor, or manager; a current or former spouse or lover; or a relative, friend, or some other person who has a dispute with an employee of the affected workplace.

Type I Events

The majority (60%) of workplace homicides involve a person entering a small late-night retail establishment, e.g., liquor store, gas station, or convenience food store, to commit a robbery. During the commission of the robbery, an employee or, more likely, the proprietor is killed or injured.

Employees or proprietors who have face-to-face contact and exchange money with the public, work late at night and into the early morning hours, and work alone or in very small numbers are at the greatest risk of a type I event. While the assailant may feign to be a customer as a pretext to enter the establishment, he/she has no legitimate business relationship to the workplace.

Type II Events

A type II workplace violence event involves an assault by someone who is either the recipient or the object of a service provided by the affected workplace or the victim. Even though type I events represent the most common type of fatality, type II events involving victims who provide services to the public are also increasing. Type II events accounted for approximately 30% of workplace homicides. Further, when more occupation-specific data about nonfatal workplace violence become available, nonfatal type II events involving assaults to service providers, especially to health care providers, may represent the most prevalent category of workplace violence resulting in physical injury. Type II events involve fatal or nonfatal injuries to individuals who provide services to the public. These events include assaults on public safety and correctional personnel, municipal bus or railway drivers, health care and social service providers, teachers, sales personnel, and other public or private service sector employees who provide professional, public safety, administrative, or business services to the public. This is also prevalent for public safety services (i.e., law enforcement).

Type III Events

A type III workplace violence event consists of an assault by an individual who has some employment-related involvement with the workplace. Generally, a type III event involves a threat of violence, or a physical act of violence resulting in a fatal or nonfatal injury to an employee, supervisor, or manager of the affected workplace by the following types of individuals:

- Current or former employee, supervisor, or manager
- Some other person who has a dispute with an employee of the affected workplace, e.g., current or former spouse or lover, relative, friend, or acquaintance

Type III events account for a much smaller proportion of fatal workplace injuries. Type III events accounted for only 10% of workplace homicides.

Why Address Workplace Violence?

If the risk of workplace violence exists, this is the reason to address security that will provide protection to an employer's workforce. The following are why it needs to be addressed:

- Workplace violence is a real issue; thus, it is a reality.
- History has shown that the potential for workplace violence exists.
- It is a real risk that could impact all workplaces.
- The statistics indicate that a number of deaths and injuries have resulted from this type of violence.
- Certain employers and their workplaces have more of a prevalence for violence.
- Certain employee occupations are more at risk of violence than others.
- Some physical workplaces are more vulnerable to violence than others.
- The presence or handling of money tends to increase the risk.
- Certain clients (mentally ill) increase the risk of violence.
- Employees who travel or work outside the employer's facilities have more potential risk or exposure.
- It is the responsibility of the employer to provide a safe workplace for his/her employees.

Summary

Effective security management to prevent all three types of workplace violence events also includes postevent measures such as emergency medical care and debriefing employees about the incident. After a workplace assault occurs, employers should provide postevent trauma counseling to those who desire such intervention to reduce the short- and long-term physical and emotional effects of the incident.

Workplace safety and health hazards affecting employees have traditionally been viewed as arising from unsafe work practices, hazardous industrial conditions, or exposures to harmful chemical, biological, or physical agents, not from violent acts committed by other human beings. Recently, though, employees, as well as supervisors and managers, have become, all too frequently, victims of assaults or other violent acts in the workplace that entail a substantial risk of physical or emotional harm. Many of these assaults result in fatal injury, but an even greater number result in nonfatal injury, or in the threat of injury, which can lead to medical treatment, missed work, lost wages, and decreased productivity.

A single explanation for the increase in workplace violence is not readily available. Some episodes of workplace violence, like robberies of small retail establishments, seem related to the larger societal problems of crime and substance abuse. Other episodes seem to arise more specifically from employment-related problems.

What can be done to prevent workplace violence? Any preventive measure must be based on a thorough understanding of the risk factors associated with the various types of workplace violence. Moreover, even though our understanding of the factors that lead to workplace violence is not perfect, sufficient information is available, which, if utilized effectively, can reduce the risk of workplace violence. However, strong management commitment, and the day-to-day involvement of managers, supervisors, employees, and labor unions, is required to reduce the risk of workplace violence.

Workplace violence has become a serious occupational health problem requiring the combined efforts of employers, employees, labor unions, government, academic researchers, and security professionals. The problem cannot be solved by government alone.

Further Readings

California Department of Labor, *Guidelines for Security and Safety of Health Care and Community Service Workers.* http://www.ca.gov. Sacramento, CA, March 1998.

United States Department of Health and Human Services: National Institute for Occupational Safety and Health, *Violence in the Workplace: Risk Factors and Prevention Strategies (CIB 57).* Washington, DC: US Department of Health and Human Services, June 1996.

United States Department of Justice's Bureau of Justice Statistics, *Annual National Crime Victimization Survey for 1992–1996.* Washington, DC: US Department of Justice, 1998.

United States Office of Personnel Management, *Dealing with Workplace Violence: A Guide for Agency Planners.* Washington, DC: US Office of Personnel Management, February 1998.

50

External Force: Terrorism

Since September 11, 2001, the reality of the damage and effects upon the workplace from terrorism has been realized by employers; it had not been a highly publicized fact that terrorism could in fact be a problem for employers. Those employers and employees who traveled and worked outside the United States were well aware of the potential hazards from it. Employers have had to make provisions for employees to protect company and business personnel and their facilities from dangers from outside the workplace that would inflict death and pain upon employees while exacting maximum damage upon the physical assets of the workplace.

All of a sudden, employers realized that terrorism was their problem and responsibility. Employees considered their safety and well-being threatened by the potential dangers posed by terrorism.

The tools of terrorism are the conventional weapons of war and assaults (guns, rockets, bombs, etc.) but also the weapons of nuclear, chemical, and biological (NBC) agents that could result in mass destruction and loss of life. No effort to deter terrorism is foolproof. But, taking no action to assess the risk, take control, and implement actions to protect employees and the workplace is not an option.

It cannot be assumed that all acts of terrorism can be prevented. Thus, employers must make some preparation for such events so that damage will be minimal and recovery from any such disaster will be rapid. This includes the following:

- Antiterrorism training
- Self-protective behavior
- Protection devices (e.g., bullet proof vest)
- Safe activities that decrease personal risk
- Security provisions
- Hardening potential targets
- Planning that is implemented and practiced

Most terrorist operations are planned and carried out with a degree of expertise. They seek to exploit the targets that are vulnerable while attempting to minimize their own risk, with the exception of suicide attackers. For the most part, terrorist acts are limited to six basic forms: bombings, assassinations, armed assaults, kidnapping, barricade and hostage situations, and hijackings. Bombings are the most common. All in their most basic form are simple criminal acts. The manner in which they are carried out, the victims that are targeted, and the desired media and political outcomes are the only differences between terrorists and common criminals.

Travel Security

When executives, managers, and employees travel or work abroad or go to areas where they have potential for being attacked, assaulted, injured, killed, or kidnapped, a standard of travel practices should be developed and followed. Each deployment or trip should be planned with employee security and any potential threat considered foremost. Employers need to consider the travelers' well-being by doing the following:

- Having standard company policies for travel as well as curfews and off-limits locations
- Having situational awareness
- Planning an itinerary (movement, dates, times, and location, even use of GPSs)
- Avoiding high-risk locations
- Encouraging travelers to maintain a low profile
- Varying routes, times, and patterns of behavior
- Providing security detail (guards, etc.)
- Providing medical assistance
- Maintaining communications
- Having evacuation plans
- Providing travel insurance for theft, identity, cash advances, travel interruptions, etc.
- Securing unique situational measures (rescue, kidnaping, etc.)

Hardening Facilities Increases Protection

Some precautions can be undertaken to help preclude the ability of a terrorist to easily access the workplace facility or to protect vulnerable areas near and around company compounds.

Some cost may be well worth the investments depending upon the risk to the workplace facility and its occupants. Some of the actions that can be undertaken are as follows:

- Install a perimeter around the compound or facility.
- Consider installing telephone caller ID; record phone calls, if necessary.
- Increase perimeter lighting.
- Deploy visible security cameras and motion sensors.
- Remove vegetation in and around perimeters; maintain regularly.
- Institute a robust vehicle inspection program to include checking the undercarriage of vehicles, under the hood, and in the trunk. Provide vehicle inspection training to security personnel.
- Deploy explosive detection devices and explosive detection canine teams. Conduct vulnerability studies focusing on physical security, structural engineering, infrastructure engineering, power, water, and air infiltration, if feasible.

- Initiate a system to enhance mail and package screening procedures (both announced and unannounced).
- Install special locking devices on manhole covers in and around facilities.
- Install intrusion devices.
- Implement a countersurveillance detection program.

Blast protection can be undertaken during the design phase to prevent and delay facility collapse. Other actions that can be taken to make facilities more blast resistant are as follows:

- Use configurations that better resist blast shockwaves.
- Maximize distances between parking and facilities.
- Review size and location of windows with detonation points in mind, and use blast- or ballistic-resistant glazing.
- Increase the strength of exterior cladding and nonstructural elements.
- Avoid exterior ornamentation that can break away.

Potential Terrorist's Weapons

Newspapers or magazines are frequently printing headlines that warn about the potential for another terrorist attack. Turn on the television, and it is broadcasting about the most recent bombing. On a daily basis, the world is confronted with the potential for another major terrorist attack; this time, the terrorist may be using NBC agents. Looking back through history, acts of biological and chemical terrorism have been happening for centuries. But seldom have they been used against the workplace and its occupants; that possibility is changing. An understanding of the potential types of weapons that terrorists could use is educational, and it is somewhat reassuring that their use can be mitigated by understanding and planning. When discussing NBC use as weapons against the workplace, it is usually in conjunction with a terrorist-style activity.

Protection from Chemical, Biological, or Radiological Attacks

There are no foolproof ways to prevent the effects of a terrorist attack on the workplace. Also, there is no way to predict which facility might be attacked. But, knowledge of the business or businesses in the vicinity may make it more likely that certain facilities will be targeted for an attack using NBC; this will help determine the risk. If particular businesses or companies are an easy target and have done nothing to act as a deterrent from an NBC attack, then employers are being negligent since it is know that such an attack is a possibility.

Everything in this section assumes that the attack on your facility will use airborne materials that can be injected into the workplace heating, ventilation, and air-conditioning (HVAC) system and dispersed throughout the facility. Understanding the systems in your facility and how they operate will allow you to incorporate some of the recommendation in this section, thus better protecting your workplace and its workforce from an NBC attack.

Preventing terrorist access to a targeted facility requires physical security of entry, storage, roof, and mechanical areas, as well as securing access to the outdoor air intakes of the facility's HVAC system. The physical security needs of each workplace should be assessed, as the threat of an NBC attack will vary considerably from workplace to workplace.

Procedures and preventive maintenance schedules should be implemented for cleaning and maintaining ventilation system components. Replacement filters, parts, and so forth should be obtained from known manufacturers and examined prior to installation. It is important that ventilation systems be maintained and cleaned according to the manufacturer's specifications.

Summary

The reality of today's world is that terrorist incidents could happen and will happen. Thus, it is unrealistic that a business or company does not take actions to prevent or at least mitigate potential damage from a terrorist event.

Reducing a facility's vulnerability to an airborne NBC attack requires a comprehensive approach. Decisions concerning which protective measures to implement should be based upon the threat profile and a security assessment of the workplace and its occupants. While physical security is the first layer of defense, other issues must also be addressed.

This is why preventing possible terrorist access to outdoor air intakes and mechanical rooms and developing NBC-contingent emergency response plans should be addressed as soon as possible. Additional measures can provide further protection. A security assessment should be done to determine the necessity of additional measures. Some items, such as improved maintenance and HVAC system controls, may also provide a payback in operating costs and/or improved workplace air quality. As new facility designs or modifications are considered, designers should consider that practical NBC sensors may soon become available. Facility system design features that are capable of incorporating this rapidly evolving technology will most likely offer a greater level of protection.

Further Readings

American Institute of Physics. *"Dirty Bombs" Much More Likely to Create Fear than Cause Cancer.* March 12, 2002.

Buck, G. *Preparing for Biological Terrorism.* Albany, NY: Delmar, 2002.

Centers for Disease Control and Prevention. *Biological and Chemical Terrorism: Strategic Plan for Preparedness and Response.* Morbidity and Mortality Weekly Report. April 21, 2000, Vol. 49/No. RR-4.

Centers for Disease Control and Prevention. *Radiation Emergencies Fact Sheet.* Available at http://emergency.cdc.gov/radiation/factsheets.asp.

National Conference of State Legislatures. *Environmental Health Series, Biological and Chemical Terrorism.* January 2003, Volume 7.

National Institute for Occupational Safety and Health. *Interim Recommendations for the Selection and Use of Protective Clothing and Respirators against Biological Agents.* October 24, 2001.

National Institute for Occupational Safety and Health. *Protecting Building Environments from Airborne Chemical, Biological, or Radiological Attacks.* May 2002.

Osterholm, M.T. *Bioterrorism: A Real Modern Threat. Emerging Infections 5.* Washington, DC: ASM Press, 2001.

Reese, C.D. *Office Building Safety and Health.* Boca Raton, FL: CRC Press, 2004.

Reese, C.D. *Occupational Health and Safety Management (Third Edition).* Boca Raton, FL: CRC Press, 2016.

51

Off-the-Job Safety

The National Safety Council in *Injury Facts* provides some insight into why off-the job safety is a topic that should receive some attention when occupational safety and health (OSH) is addressed. For instance, over 50% of the accidents that keep employees off the job occur away from work. Employers spend in excess of $500 per employee to cover health care costs for employees and their families for off-the-job accidents/incidents. Also, motor vehicles are the leading cause of death, and this is especially true for teens and younger individuals. Over 15 million individuals suffer temporary or permanent disability injuries from off-the-job injuries each year. Accidents are the leading cause of death in those below the age of 65. The facts help set the stage as to why off-the-job safety is a topic that deserves attention.

It might be wondered why employers would bother themselves with safety off the job. There is still cost to the employer when accidents to the workers or their families occur during time away from the workplace. The aftermath of an off-the-job event can be disruptive to the family unit in many ways. These include stress to the family unit as well as absences from the workplace during and after a family incident, even if it not the worker personally.

It is difficult for an employer to mandate a value of safety and health when an employee and his/her family are not in the workplace. Encouraging employees and their families to place a value on both safety and health away from the workplace is often a difficult sale to employees, who feel that their private time is none of the employer's business.

Off-the-job incidents can cost the family physically as well as damage their personal property such as homes, automobiles, boats, motorcycles, etc. Incidents may disrupt the normal functioning of the family if the employee or spouse is disabled. The loss of a second income or the need for childcare may place a financial burden in such events. If a child is disabled, a parent may need to stay with him/her. This again may place a financial burden on the family unit.

The difficulties faced by an employer may be potential increases in premiums for family medical insurance, loss of a productive employee, and loss of production when using a temporary employee. When an employee suffers an off-the-job injury or health issue, very often, it is disabling.

The types of accidents suffered off the job are often similar to those that occur in the workplace. Employees often suffer injuries during risky or dangerous hobbies, pleasure, or sport activities. Most often, they are the types of hazards found in the workplace, such as the following:

- Automobile accidents
- Fires and burns
- Motorized vehicles such as ATVs, boats, motorcycles, lawn tractors, etc.
- Bicycles
- Slips, trips, or falls

- Falls from elevated surfaces such as roofs and ladders
- Lawn mowing and trimming equipment
- Power and hand tools
- Poisoning
- Hazardous chemicals
- Drowning
- Firearms

Companies may want to develop safety and health materials to distribute as paycheck stuffers that are seasonally related, such as safe skiing tips in winter or reminders on flu shots during flu season. More extensive programs may be implemented for disease prevention emphasizing vaccinations, hygiene practices, healthy eating habits, and prevention planning.

Since automobile accidents transpire often, the company may offer a safe driving course for the family or especially oriented toward teen drivers since they are most often the least experienced members of the family.

Also, information on emergency planning, evacuation, emergency numbers, and first aid might be provided. Companies might distribute packages for emergency procedures or first-aid kits.

These items can be developed as the company develops a prevention approach and plan for off-the-job safety. This is why an off-the-job safety and health plan needs to be developed that counts the number and types of off-the-job injuries and illnesses the workforce is experiencing, and the medical cost and indirect cost (lost wages, cost other than medical, etc.) of such injuries and illnesses. Emphasis on prevention should be placed upon frequent and severe occurrences. Special materials should be developed by the company to assist employees and their families in preventing these types of incidents. Track the gains toward goals of prevention and cost savings to the employer and employees.

Further Readings

Goetsch, D.L. *Occupational Safety and Health for Technologists, Engineers and Managers (Fifth Edition)*. Upper Saddle River, NJ: Pearson Prentice Hall, 2005.

Kohn, J.P. and T.S. Ferry. *Safety and Health Management Planning*. Rockville, MD: Government Institutes, 1999.

Lack, R.W. *Safety, Health, and Asset Protection: Management Essential (Second Edition)*. Boca Raton, FL: Lewis Publishers, 2002.

Section K

Miscellaneous Safety and Health Factors

At certain times, some topics with relevance to occupational safety and health (OSH) do not fit neatly into a definitive category, but nonetheless, they are important to workers and workplace safety and health and must somehow fit under an odd category, such as "miscellaneous."

The contents of this section are as follows:

Chapter 52—Human Factors
Chapter 53—Ergonomics
Chapter 54—Product Safety

52

Human Factors

Defining Human Factors

When discussing occupational safety and health (OSH), it seems only logical to discuss how the interaction between machines and humans has an impact upon occurrences of accidents/incidents in a workplace. It is unrealistic to discuss or incorporate OSH into the work environment without taking into account where humans fit into the puzzle.

With this said, addressing human factors is not a single-discipline endeavor. Multiple fields of science and their data and principles must be an integral part of this undertaking. These science disciplines include but are not limited to engineering, sociology, anthropology, medicine, anatomy, psychology, physiology, biology, computer sciences, etc.

Human factors must pull together research data, practical knowledge, approaches used by other industries, and safety and health as applied to human–machine interfaces. In order to be effective, human factors must be there from the beginning (predesign analysis) as to the task to be accomplished. Secondly, they must be present during detailed engineering design, including environmental factors, operator function and interfaces, and reduction of operator dangers and stressors to design a more efficient/productive system with reduced risk, strain, and stress on operators that improves operator/machine compatibility. Achieving this cannot be assumed from the design but must be tested and evaluated to assure that design goals have been achieved.

Human Factor Safety

The ultimate goal would be to design out the inherent hazards or risk. This would consider in some detail the human effort model since humans, by their very presence in the system, interject causal factors. These causal factors go beyond deciding to make mistakes to the limiting factors inherent to humans such as fatigue, physical limits, intellectual factors, etc.

Everything that requires human interaction must be designed around human limits and limitations since there are certain tasks that machines do better than humans, such as constant-pressure or constant-force tasks or fast-repeating tasks over sustained periods. While human limits related to strength, climate, illumination, and visual acuity may be limited, humans have greater flexibility related to mental capacity by being able to change responses, make informed decisions, and take corrective actions.

But humans often require extensive training as the task/response requirements increase in detail or complexity.

Human Characteristics

The human side of human factors involves the task of fitting people into the workplace in a safe and healthy manner. This is most recognizable when the statement is made that "people are different." This is seen in the size, length, and width of people, which do not permit a one-size-fits-all answer for issues of trying develop workplaces, furniture, tools, and equipment to accommodate all. This has resulted in most items being adjustable, but the extremes may still not fit. When reworking a workplace is not possible, then limitations on hiring may have to be implemented, and now that becomes a human resource issue.

Also, humans have been conditioned over time to expect certain actions to work in a predictable way, and if they do not, this could cause an accident or incident. If we want someone to think before equipment is turned off, the turning-off direction threading would be reversed from clockwise to counterclockwise. To slow a car's speed, the brake would be applied by pushing on the pedal. This is the expected human response to stopping.

Why Human Causal Factors?

For many years, it has been thought that humans were a leading causal agent of accidents/ incidents within the workplace. Since this topic has been researched widely, there are a number of theories as to why it has some validity and even a specific topic that addresses the human factor in the field of occupational safety and health.

Just like machines, humans can experience overload from tasks beyond their physical and mental capabilities that can transpire due to the environment (i.e., temperature extremes), internal factors (i.e., personal/emotional issues), or situational factors (i.e., job expectations or risk levels). Humans make wrong types of decisions such as ignoring safety or removing safeguards. Also, humans often undertake tasks without having the need skill or understanding the risk involved.

Benefits of Using Human Factors

The benefits from using human factors are as follows:

- It improves safety and health.
- It increases quality.
- It increases productivity.
- Poor designs are costly.
- It fits the work to the person.
- It makes work more user friendly.
- It can be used by all businesses.
- It is a tool for safety and health management.

Summary

Applying human factor principles means the following:

- People are very important.
- Concern for employee well-being is important.
- As with any good management approach, workers must be involved and empowered.
- It improves morale and enhances employee and labor relations.
- It positions the business for changing trends and improves potential profitability.
- It improves occupational safety and health.
- It provides flexibility in responding to all, leading to a competitive business edge.
- It allows for quicker problem solving.

When the discussion concerns human factors, it becomes a philosophical back and forth, as it is really ergonomics. Earth looks at the two (human factors and ergonomics) as one and the same. So for simplicity purposes, just consider them synonymous, since their common purpose is to make the workplace safer and healthier by making the interface between the work environment, its machine/equipment, and its humans (employees) a safer and healthier place to work. Is this not the ultimate goal?

Further Readings

Bahr, N.J. *System Safety Engineering and Risk Assessment: A Practical Approach.* New York: Taylor & Francis, 1997.

Burns, T.E. *Serious Incident Prevention (Second Edition).* Boston, MA: Gulf Professional Publishing, 1946.

Daugherty, J.E. *Industrial Safety Management: A Practical Approach.* Rockville, MD: Government Institutes, 1999.

Goetsch, D.L. *Occupational Safety and Health for Technologists, Engineers, and Managers (Fifth Edition).* Upper Saddle River, NJ: Pearson Prentice Hall, 2005.

Kohn, J.P. and T.S. Ferry. *Safety and Health Management Planning.* Rockville, MD: Government Institutes, 1999.

Lack, R.W. *Safety, Health, and Asset Protection: Management Essentials (Second Edition).* Boca Raton, FL: Lewis Publishers, 2002.

MacLeod, D. *The Ergonomics Edge: Improving Safety, Quality, and Productivity.* New York: Van Nostrand Reinhold, 1995.

McCormick, E.J. *Human Factors in Engineering and Design.* New York: McGraw-Hill, 1976.

53

Ergonomics

When the word *ergonomics* is mentioned, most employers and many others will tend to make derogatory comments and act as though ergonomics is some kind of contrived problem. However, from experience of visiting many types of industries and workplaces, it is not at all unusual to talk to a person who has had five surgeries related to musculoskeletal disorders (MSDs), cumulative trauma disorders (CTDs), repetitive strain injuries, sprains/strains, or repetitive-motion injuries (RMIs). In this chapter, these terms will be used interchangeably.

Since the Bureau of Labor Statistics (BLS) reported ergonomic-related injuries/disorders as illnesses until 2012, even though they made up approximately 69% of reported illnesses, now such injuries/illnesses are reported as "other" without recognizing that they have an origin in ergonomics. Because there is not one word that describes ergonomic-related injuries or illnesses, they are described as a mixed bag of conditions that, in most cases, require medical treatment and quickly become costly and disabling.

As workers conduct the tasks of their jobs, there is a constant wear and tear on their bodies. Ergonomic risk factors are the aspects of a job or task that impose a biomechanical stress on the worker.

The factors that exacerbate these types of injuries/illnesses are as follows:

- Force
- Vibration
- Repetition
- Contact stress
- Awkward postures
- Cold temperatures
- Static postures

A brief description of why these factors are important to addressing and solving ergonomic issues is as follows.

Force

Force refers to the amount of physical effort that is required to accomplish a task or motion. Tasks or motions that require application of higher force place higher mechanical loads on muscles, tendons, ligaments, and joints. Tasks involving high forces may cause muscles to fatigue more quickly. High forces may also lead to irritation, inflammation, strains, and tears of muscles, tendons, and other tissues.

Repetition

Repetition refers to performing a task or series of motions over and over again with little variation. When motions are repeated frequently (e.g., every few seconds) for prolonged periods (e.g., several hours, a work shift), fatigue and strain of the muscle and tendons can occur because there may be inadequate time for recovery. Repetition often involves the use of only a few muscles and body parts, which can become extremely fatigued, while the rest of the body works very little.

Awkward Postures

Awkward postures refer to positions of the body (e.g., limbs, joints, back) that deviate significantly from the neutral position while job tasks are performed. For example, when a person's arm is hanging straight down (i.e., perpendicular to the ground) with the elbow close to the body, the shoulder is said to be in a neutral position. However, when employees are performing overhead work (e.g., installing or repairing equipment, grasping objects from a high shelf), their shoulders are far from the neutral position.

Static Postures

Static postures (or static loading) refer to physical exertion in which the same posture or position is held throughout the exertion. These types of exertions put increased loads or forces on the muscles and tendons, which contribute to fatigue. This occurs because not moving impedes the flow of blood that is needed to bring nutrients to the muscles and to carry away the waste products of muscle metabolism.

Vibration

Vibration is the oscillatory motion of a physical body. Localized vibration, such as vibration of the hand and arm, occurs when a specific part of the body comes into contact with vibrating objects such as powered hand tools (e.g., chain saw, electric drill, chipping hammer) or equipment (e.g., wood planer, punch press, packaging machine). Whole-body vibration occurs when standing or sitting in vibrating environments (e.g., driving a truck over bumpy roads) or when using heavy vibrating equipment that requires whole-body involvement (e.g., jackhammers).

Contact Stress

Contact stress results from occasional, repeated, or continuous contact between sensitive body tissue and a hard or sharp object. Contact stress commonly affects the soft tissue on the fingers, palms, forearms, thighs, shins, and feet. This contact may create pressure over a small area of the body (e.g., wrist, forearm), which can inhibit blood flow, tendon and muscle movement, and nerve function. Examples of contact stress include resting wrists on the sharp edge of a desk or workstation while performing tasks and the pressing of tool handles into the palms.

Cold Temperatures

Cold temperatures refer to exposure to excessive cold while performing work tasks. Cold temperatures can reduce the dexterity and sensitivity of the hand. Cold temperatures, for example, cause the worker to apply more grip force to hold hand tools and objects. Also, prolonged contact with cold surfaces (e.g., handling cold meat) can impair dexterity and induce numbness. Cold is a problem when it is present with other risk factors and is especially problematic when it is present with vibration exposure.

Ergonomics-related conditions are observed in shipyards, on construction sites, in manufacturing, in the service industry, and in the office environment. When ergonomics is mentioned, many individuals immediately think of computer workstations, which are a small part of this issue, and in most cases, the problems with them are easily fixable.

Ergonomics is, by definition, fitting the workplace to the worker. It means more than changing a workstation. It means that the whole environment is designed to fit workers, including directions, controls, printed material, warning signals, mental stress, work schedules, the work climate, fatigue and boredom, material handling, noise, vibration, lighting, mental capacity, the worker–machine interface, and the list could go on.

Ergonomics brings to bear many different academic disciplines. This is especially true of the more complex workplace problems. For the most part, many solutions can be achieved simply and with little cost involved. To solve most of the problems faced with ergonomic implications, being a rocket scientist is not a requirement. The workers themselves often have very viable solutions. This is why the Occupational Safety and Health Administration (OSHA) was requiring worker involvement in their now defunct ergonomics regulations. This is not to say that some of the existing ergonomic issues in the workplace will not require some time and cost investment by the employer. In most cases, this investment in solving workplace ergonomic problems decreases injuries and improves efficiency and morale.

There needs to be a standard definition that fully describes the types of incidents related to ergonomics, which is a intermingling of a multitude of disciplines such as physiological, psychological, behavioral, and psychosocial, as an interaction of machinery/equipment, workplace design, and the human interface (workers).

Ergonomic issues have escalated in recent years, as well as the cost of such injury or illness outcomes. No industry is exempt from ergonomic issues, from the office to ship-building. Thus, companies should recognize, evaluate, analyze, plan, and implement an ergonomic program to accomplish the following:

- Mitigate ergonomic injuries/illnesses
- Assess the multitude and magnitude of the problem
- Implement controls
- Undertake workplace and workstation redesign
- Foster medical surveillance and early detection
- Conduct training and education
- Evaluate progress and goal attainment

Extent of the Problem

MSDs are often referred to as ergonomic injuries or illnesses affecting connective tissues of the body such as muscles, nerves, tendons, joints, cartilage, or spinal disks.

In 2011, MSDs accounted for 387,820 cases (33%) of injuries and illnesses. Six occupations accounted for 26% of the total. They are nursing assistants, laborers, janitor/cleaners, heavy and light tractor-trailer truck drivers, registered nurses, and stock clerks.

MSDs account for 335,390 (33%) of all workplace injuries requiring time away from work as reported by the BLS in 2007, compared with 30% in 2006. Nursing aides and orderlies had the highest rate of injury, followed by emergency medical personnel, laborers/material movers, ticket agents/travel clerks, light truck drivers, and heavy truck drivers. Decreases in MSDs were seen in company management, construction, and manufacturing.

In 2004, the BLS reported 402,700 MSDs. Of these, sprains, strains, and tears accounted for more than three-fourths of the MSDs that resulted in days away from work. Service-providing industries reported the most MSDs, accounting for 69% of all cases. Within these industries, health care and social assistance reported the most cases. Goods-producing industries reported 31% of all MSD cases, led by manufacturing with about 20%. The three main occupations with the most MSDs were laborers and material movers; nursing aides, orderlies, and attendants; and heavy and tractor-trailer truck drivers.

Developing an Ergonomic Program

When ergonomic-related issues exist, it would be suggested that the first undertaking be to develop a written ergonomic incident prevention program. This program should contain the elements that will lead to prevention. This would include management commitment and employee involvement. (Employee involvement is critical in solving ergonomic-related problems.) The second element should be hazard identification and assessment. The third element would be hazard control and prevention, managing injuries/illnesses, and education and training. Lastly, evaluate its effectiveness and progress.

An ergonomics program that contains well-recognized program elements is necessary. All of these core elements are essential to the effective functioning of ergonomics programs. These elements have been endorsed by both private industry and OSHA as keys to ergonomic program effectiveness. All of the elements are important, although many safety and health professionals believe that management leadership and employee participation are the keystone of an effective ergonomics program.

Physical Work Activities and Conditions

The following physical work activities and conditions contribute to ergonomic potential for ergonomic-related problems:

- Physical demands of work
- Workplace and workstation conditions and layout
- Characteristics of objects that are handled or used
- Environmental conditions

Employers should examine a job in which an ergonomic injury/illness has occurred to identify the physical work activities and workplace conditions and then evaluate the risk factors to assess the work environment. This will help to limit exposures.

Limits of Exposure

To make a determination as to the real risk, the following should be examined: duration, frequency, and magnitude (i.e., modifying factors) of the employee's exposure to the ergonomic risk factors. These risk factors do not always rise to the level that poses a significant risk of injury. This may be because the exposure does not last long enough, is not repeated frequently enough, or is not intensive enough to pose a risk.

Ergonomic Controls

Controls that reduce a risk factor focus on reductions in the risk modifiers (frequency, duration, or magnitude). By limiting exposure to the modifiers, the risk of an injury is reduced. Thus, in any job, the combination of the task, environment, and worker create a continuum of opportunity to reduce the risk by reducing the modifying factors. The closer the control approach comes to eliminating the frequency, duration, or magnitude, the more likely it is that the ergonomic hazard has been controlled. Conversely, if the control does little to change the frequency, duration, or magnitude, it is unlikely that the ergonomic hazard has been controlled.

In determining control, ask employees in the problem job for recommendations about eliminating or materially reducing the ergonomic hazards. Second, identify, assess, and implement feasible controls (interim and permanent) to eliminate or materially reduce the ergonomic hazards. This includes prioritizing the control of hazards, where necessary. Thirdly, track your progress in eliminating or materially reducing these hazards. This includes consulting with employees in problem jobs about whether the implemented controls have eliminated or materially reduced the hazard. And last, identify and evaluate ergonomic hazards when you change, design, or purchase equipment or processes in problem jobs.

Identifying Controls

There are different methods that can be used and places to consult to identify controls. Many employers rely on their internal resources to identify possible controls. These in-house experts may include the following:

- Employees who perform the job and their supervisors
- Engineering personnel
- Workplace safety and health personnel or committee
- Maintenance personnel
- On-site health care professionals
- Procurement staff
- Human resource personnel

Possible controls can also be identified from sources outside the workplace, such as the following:

- Equipment catalogs
- Vendors
- Trade associations or labor unions
- Conferences and trade shows
- Insurance companies
- OSHA consultation services
- Specialists

Assessing Controls

The assessment of controls is an effort by management, with input from employees, to select controls that are reasonably anticipated to eliminate or materially reduce the ergonomic hazards. Several controls may be available that would be reasonably likely to reduce the hazard. Multiple control alternatives are often available, especially when several risk factors contribute to the ergonomic hazard. The employer may need to assess which of the possible controls should be tried. Clearly, a control that significantly reduces several risk factors is preferred over a control that only reduces one of the risk factors.

Selection of the risk factor(s) to control, and control measures to try, can be based on numerous criteria. An example of one method involves ranking all of the ergonomic risk factors and possible controls according to how well they meet these four criteria:

- Effectiveness—greatest reduction in exposure to the ergonomic hazards
- Acceptability—employees most likely to accept and use this control
- Timeliness—takes least amount of time to implement, train, and achieve material reduction in exposure to ergonomic hazards
- Cost—elimination or material reduction of exposure to ergonomic hazards at the lowest cost

Implementing Controls

Because of the multifactorial nature of MSD hazards, it is not always clear whether the selected controls will achieve the intended reduction in exposure to the hazards. As a result, the control of ergonomic hazards often requires testing selected controls and modifying them appropriately before implementing them throughout the job. Testing controls verifies that the proposed solution actually works and indicates what additional changes or enhancements are needed.

Tracking Progress

First, evaluating the effectiveness of controls is top priority in an incremental abatement process. Unless they follow up on their control efforts, employers will not know whether the hazards have been adequately controlled or whether the abatement process needs to continue. That is, if the job is not controlled, the problem solving is not complete.

Second, the tracking of progress is also essential in cases where a need exists to prioritize the control of hazards. It denotes whether the abatement plans are on schedule.

Third, tracking the progress of control efforts is a good way of determining whether the elements of the program are functioning properly and quantifying their success. The following are some of the measures to use:

- Reduction in severity rates, especially at the very start of the program
- Reduction in incidence rates
- Reduction in total lost workdays and lost workdays per case

- Reduction in job turnover or absenteeism
- Reduction in workers' compensation costs and medical costs
- Increases in productivity or quality
- Reduction in reject rates
- Number of jobs analyzed and controlled
- Number of problems solved

Proactive Ergonomics

Sometimes this approach is referred to as proactive ergonomics or safety through design. The concept encompasses facilities, hardware, equipment, tooling, materials, layout and configuration, energy controls, and environmental concerns and products. Designing or purchasing to eliminate or materially reduce ergonomic hazards in the design process helps to avoid costly retrofitting. It also results in easier and less costly implementation of ergonomic controls.

Ergonomists endorse the hierarchy of controls, which accords first place to engineering controls, because these control technologies should be selected based on their reliability and efficacy in eliminating or reducing the workplace hazard (risk factors) giving rise to the ergonomic issues. Engineering controls are preferred because these controls and their effectiveness have the following characteristics:

- Reliable
- Consistent
- Effective
- Measurable
- Not dependent on human behavior (that of managers, supervisors, or workers) for their effectiveness
- Do not introduce new hazards into the process

In contrast to administrative and work practice controls or personal protective equipment, which occupy the second and third tiers of the hierarchy, respectively, engineering controls fix the problem once and for all. However, because there is such variability in workplace conditions, there is a need to use any combination of engineering, work practice, or administrative controls as methods of control for ergonomic hazards.

Education and training can be used in a variety of ways. The foremost is to train all employees in ergonomic hazard awareness, the program and procedures, sign and symptom identification, and types of injuries and illnesses. Second, train some of the workforce in ergonomic assessment so that there will be teams of both management and labor to evaluate ergonomic hazards and make recommendations for controlling the potential risk factors on the jobs in your workplace. With proper training, you can have an educated workforce who can be assets rather than liabilities in solving ergonomic problems.

Ergonomic principles are most effectively applied to workstations and new designs on a preventive basis, before injuries or illnesses occur. Good design with ergonomics provides the greatest economic benefit for industry. Design strategies should emphasize fitting job demands to the capabilities and limitations of employees. To achieve this, decision makers must have appropriate information and knowledge about ergonomic risk factors and ways to control them. They need to know about the problems in jobs and the causes. Designers of in-house equipment, machines, and processes also need to have an understanding of ergonomic risk factors and know how to control them. For example, they may need anthropometric data to be able to design to the range of capabilities and limitations of employees.

It is also important that persons involved in procurement have basic knowledge about the causes of problems and ergonomic solutions. For example, they need to know that adjustable chairs can reduce awkward postures and that narrow tool handles can considerably increase the amount of force required to perform a task. In addition, to prevent the introduction of new hazards into the workplace, procurement personnel need information about equipment needs.

Ergonomics is a continuous improvement process. If an employer can show that they have made an organized effort to identify ergonomic stressors, to educate affected employees on ergonomic principles, to implement solutions, and to have a system to identify when a solution is not working and needs to be readdressed, then a giant step has been taken toward mitigating your ergonomic problems.

Further Readings

California Department of Industrial Relations (Cal/OSHA), *Easy Ergonomics: A Practical Approach for Improving the Workplace.* 1999.

Reese, C.D. *Accident/Incident Prevention Techniques (Second Edition).* Boca Raton, FL: CRC Press, 2012.

Reese, C.D. and J.V. Eidson. *Handbook of OSHA Construction Safety & Health (Second Edition).* Boca Raton, FL: CRC/Taylor & Francis, 2006.

United States Department of Health and Human Services: National Institute for Occupational Safety and Health, *Elements of Ergonomics Programs (DHHS-97–117).* 1997.

United States Department of Labor. Occupational Safety and Health Administration. *Job Hazard Analysis and Control (1910.917–922),* Subject Index. "Internet" 2001. Available at http://www.osha.gov.

United States Department of Labor. Occupational Safety and Health Administration. Subject Index. "Internet." April 1999. Available at http://www.osha.gov.

54

Product Safety

The quality of the product is primary to good business, as is whether it meets the goals of its intended use. If used as intended, the product should fulfill its purpose and with an expectation of being safe. A safe product enhances the company's credibility unless flaws in the design and engineering process resulted in a product with concerns.

Product safety issues either in design or in manufacturing can result in product liability depending upon the nature and severity of the outcomes of the product's failures. Product liability impacts the company's reputation as well as decreases sales of the product. These types of product safety issues will result in cost to redesign and retooling to manufacture a safe product. The cost of an unsafe or poor-quality product will have a huge impact upon the company's bottom line.

A system safety program should exist from concept, through design, production, manufacture, distribution, marketing, and customer relations, to disposal.

Why a Safe Design Approach Should Be Employed

In the design of a safe product, consider this:

- It is cheaper to eliminate occupational safety and health hazards by well-informed decisions in the design or planning stage than when the hazards become real risks to customers or workers.
- By adopting a safe design approach, it is possible to design out safety and health hazards to create a design option to meet the end users' needs and meet the obligations under workplace safety and health legislation. A safer product will result in the benefit of a good business model, now and in the long term.

Principles of Safe Design

1. There is an understanding of the safety and health requirement of the design.
2. There is systematic hazard identification and risk evaluation.
3. Interaction occurs between people involved in the life cycle of the designed product.
4. Contractual arrangements and procurement systems operate to minimize purchased work safety and health risk.
5. A sustainable designed product results.

Legal Obligations

Obligations will vary depending on the relevant work safety and health legislation applicable to the designed product or business. The safe design approach provides for a systematic design approach. Adopting a systematic approach to work safety and health risk management can help to meet legal obligations.

Consultation

Safe design supports a collaborative risk management approach. This means that, in the design stage, people with knowledge of each phase of the product—*from design to use to dismantling*—tell the decision maker what the practical problems and benefits are. At the planning stage, design changes are more efficient and cheaper to implement, and can prevent real hazards coming into a workplace to cause risk of harm or injury.

Communicating residual risk from design is an important principle of safe design.

When Developing a Safe Design Action Plan

It impacts a large array of individuals such as the following

- Employees
- Designers: architects, engineers, industrial designers
- Manufacturers, importers, suppliers
- Employers
- Educators
- Jurisdictions
- Work safety and health practitioners
- Consumers

What Is Consumer Product Safety Screening?

Product safety screening is necessary to protect consumers from harm when using commercially available personal care products and household or workplace chemicals.

Product safety screening also provides physicians and veterinarians with important treatment information in the event of accidental poisoning of people or animals.

Why Are Products Screened for Safety?

Consumers expect that the products they use in everyday life are safe when used as directed. Product manufacturers have a moral and legal obligation to screen products and protect consumers from harm.

How Is Safety Screening Done?

Researchers use a variety of methods, including nonanimal systems (cell and tissue cultures, mathematical models, and computer models), laboratory animals, and clinical testing with human volunteers, to ensure product safety.

Like pieces of a puzzle, each type of testing is interrelated and may provide important information to help researchers piece together a complete picture of the safety of a new ingredient or product.

Labeling—What Does It All Mean?

Consumer product labeling is often misleading. If a product is sold and its safety is not known, federal regulation requires specific labeling with the following statement: "WARNING: The safety of this product has not been determined."

Marketing materials should not be misleading, inaccurate, or "use only as intended," or contain a disclaimer, servicing information, or contact information.

Summary

In summary, the process of developing a safe product carries with it a commitment for cost at each stage of development. The cost of designing a new product usually includes safety hazard analysis, tooling, manufacturing, marketing, distribution, product liability, and extra hired expertise as needed to assure a safe product. The cost of an unsafe product can and often is more expensive in the long run. These costs are also bad for business. They are as follows:

- Loss of credibility
- Cost of recalls
- Loss of vendors
- Loss of customers
- Retooling cost

- Remanufacturing cost
- Decreased profits
- Cost of legal expertise
- Cost of technical expertise

This could be avoided by initial development of a safe product.

Further Readings

Goetsch, D.L. *Occupational Safety and Health for Technologists, Engineers and Managers (Fifth Edition)*. Upper Saddle River, NJ: Pearson Prentice Hall, 2005.

Kohn, J.P. and T.S. Ferry. *Safety and Health Management Planning*. Rockville, MD: Government Institutes, 1999.

Lack, R.W. *Safety, Health, and Asset Protection: Management Essentials (Second Edition)*. Boca Raton, FL: Lewis Publishers, 2002.

Section L

Safe Facilities and Handling

Workers and the public expect that workplaces are safe for occupancy and precautions have been taken to prevent fires. The expectation is that workers and the public are protected from the hazards of materials that are manufactured and that any hazardous waste that exists is stored in a safe manner and moved around the facility in cautious manner. Even when these are transported to other facilities or across the country, the same steps are taken to protect the workforce and the public, who have potential for exposure in some manner.

The contents of this section are as follows:

Chapter 55—Fire Prevention and Life Safety

Chapter 56—Hazardous Materials

Chapter 57—Transportation

55

Fire Prevention and Life Safety

Why Address Fire Hazards?

Workplace fires and explosions kill 200 and injure more than 5,000 workers each year. In 1995, more than 75,000 workplace fires cost businesses more than $2.3 billion. Fires wreak havoc among workers and their families and destroy thousands of businesses each year, putting people out of work and severely impacting their livelihoods. The human and financial toll underscores the serious nature of workplace fires.

Almost everything will burn within the workplace. Thus, when conditions are right for fire, it can be disastrous to the workplace's physical facility, as well as to the workers, resulting in injury or death.

In order to prevent the risk of a fire, sources of fuels (burnable objects) need to be identified: the always-present atmospheric oxygen and a source of ignition such as a high temperature.

Causes of Fires

The most common causes of workplace fires are as follows:

- Electrical causes—lax maintenance in wiring, motors, switches, lamps, and heating elements
- Smoking—near flammable liquids, stored combustibles, etc.
- Cutting and welding—highly dangerous in areas where sparks can ignite combustibles
- Hot surfaces—exposure of combustibles to furnaces, hot ducts or flues, electric lamps or heating elements, and hot metal
- Overheated materials—abnormal process temperatures, materials in dryers, overheating of flammable liquids
- Open flames—gasoline or other torches, gas or oil burners
- Friction—hot bearings, misaligned or broken machine parts, choking or jamming materials, poor adjustment of moving parts
- Unknown substances—unexpected materials

323

- Spontaneous heating-deposits in ducts and flues, low-grade storage, scrap waste, oily waste and rubbish
- Combustion sparks—burning rubbish; foundry cupolas, furnaces, and fireboxes
- Miscellaneous—including incendiary cases, fires spreading from adjoining buildings, molten metal or glass, static electricity near flammable liquids, chemical action, and lighting

What the Occupational Safety and Health Administration Standards Require

Occupational Safety and Health Administration (OSHA) standards require employers to provide proper exits, firefighting equipment, and employee training to prevent fire deaths and injuries in the workplace. Each workplace building must have at least two means of escape remote from each other to be used in a fire emergency. Fire doors must not be blocked or locked to prevent emergency use when employees are within the buildings. Delayed opening of fire doors is permitted when an approved alarm system is integrated into the fire door design. Exit routes from buildings must be clear and free of obstructions and properly marked with signs designating exits from the building.

Each workplace or building must have a full complement of the proper type of fire extinguisher for the fire hazards present, excepting when employers wish to have employees evacuate instead of fighting small fires. Employees expected or anticipated to use fire extinguishers must be instructed on the hazards of fighting fire, how to properly operate the fire extinguishers available, and what procedures to follow in alerting others to the fire emergency. Only approved fire extinguishers are permitted to be used in workplaces, and they must be kept in good operating condition.

Where the employer wishes to evacuate employees instead of having them fight small fires, there must be written emergency plans and employee training for proper evacuation. Emergency action plans are required to describe the routes to use and procedures to be followed by employees. Also, procedures for accounting for all evacuated employees must be part of the plan. The written plan must be available for employee review. Where needed, special procedures for helping physically impaired employees must be addressed in the plan; also, the plan must include procedures for those employees who must remain behind temporarily to shut down critical plant equipment before they evacuate.

The preferred means of alerting employees to a fire emergency must be part of the plan, and an employee alarm system must be available throughout the workplace complex and must be used for emergency alerting for evacuation. The alarm system may be voice communication or sound signals such as bells, whistles, or horns. Employees must know the evacuation signal.

Training of all employees in what is to be done in an emergency is required. Employers must review the plan with newly assigned employees, so they know the correct actions in an emergency, and with all employees when the plan is changed.

Employers need to implement a written fire prevention plan (FPP) to complement the fire evacuation plan to minimize the frequency of evacuation. Stopping unwanted fires from occurring is the most efficient way to handle them. The written plan shall be available for employee review. Housekeeping procedures for storage and cleanup of flammable materials and flammable waste must be included in the plan. Recycling of flammable waste such as paper is encouraged; however, handling and packaging procedures must be included in the plan. Procedures for controlling workplace ignition sources such as smoking, welding, and burning must be addressed in the plan. Heat-producing equipment such as burners, heat exchangers, boilers, ovens, stoves, fryers, etc., must be properly maintained and kept clean of accumulations of flammable residues; flammables are not to be stored close to these pieces of equipment. All employees are to be apprised of the potential fire hazards of their job and the procedures called for in the employer's FPP. The plan shall be reviewed with all new employees when they begin their job and with all employees when the plan is changed. The minimum provisions that make up an FPP are as follows:

- List of all major fire hazards, proper handling and storage procedures for hazard-ous materials, potential ignition sources and their control, and type of fire protec-tion equipment necessary to control each major hazard
- Procedures to control accumulation of flammable and combustible materials
- Procedure for regular maintenance of safeguards installed on heat-producing equipment to prevent the accidental ignition of combustible materials
- Name or job title of employees responsible for maintaining equipment or control sources of ignition or fires
- Name or job title of employees responsible for the control of fuel source hazards

Any employee assigned to a job must be informed of the fire hazards to which he/she could be exposed. The employee must have received an explanation of the FPP and how it was designed to protect him/her.

Avoiding Fires

General safety precautions for avoiding fires caused by smoking are obeying "no smok-ing" signs; watching for danger spots even if no warning is posted (e.g., temporary storage area that contains combustibles); not placing lighted cigarettes on wooden tables or work-benches, even if smoking is permitted; and not putting ashes in a wastebasket or trash can.

Flammable and combustible liquids can cause fires if they are near open flames and motors that might spark. When you transfer them, bond the containers to each other and ground the one being dispensed from, to prevent sparks from static electricity. Clean up spills right away, and put oily rags in a tightly covered metal container. Change clothes immediately if you get oil or solvents on them. Watch out for empty containers that held flammable or combustible liquids; vapors might still be present. Store these liquids in approved containers in well-ventilated areas away from heat and sparks. Be sure all con-tainers for flammable and combustible liquids are clearly and correctly labeled.

Electricity can cause fires if frayed insulation and damaged plugs on power cords or extension cords are not fixed or discarded. Also, electrical conductors should not be damp or wet, and there should be no oil and grease on any wires.

A cord that's warm to the touch when current is passing through should warn you of a possible overload or hidden damage. Don't overload motors; watch for broken or oil-soaked insulation, excessive vibration, or sparks; keep motors lubricated to prevent overheating. Defective wiring, switches, and batteries on vehicles should be replaced immediately.

Electric lamps need bulb guards to prevent contact with combustibles and to help protect the bulbs from breakage. Don't try to fix electrical equipment yourself if you're not a qualified electrician.

Housekeeping is often a factor in fires in the workplace. Keep your work areas clean. Passageways and fire doors should be kept clear and unobstructed. Materials must not obstruct sprinkler heads or be piled around fire extinguisher locations or sprinkler controls. Combustible materials should be present in work areas only in quantities required for the job and should be removed to a designated storage area at the end of each workday.

Hot work such as welding and cutting should never be permitted without supervision or a hot work permit. Watch out for molten metal; it can ignite combustibles or fall into cracks and start a fire that might not erupt until hours after the work is done. Portable cutting and welding equipment is often used where it's unsafe; keep combustibles at a safe distance from hot work areas. Be sure tanks and other containers that have held flammable or combustible liquids are completely neutralized and purged before you do any hot work on them. Have a fire watch (another employee) on hand to put out a fire before it can get out of control.

Fire Safety and Protection

To protect workplaces from fire, the following items should be adhered to:

- Access to all available firefighting equipment will be maintained at all times.
- Firefighting equipment will be inspected periodically and maintained in operating condition. Defective or exhausted equipment must be replaced immediately.
- All firefighting equipment will be conspicuously located at each jobsite.
- Fire extinguishers, rated not less than 2A, will be provided for each 3,000 square feet of the protected work area. Travel distance from any point of the protected area to the nearest fire extinguisher must not exceed 100 feet. One 55-gallon open drum of water, with two fire pails, may be substituted for a fire extinguisher having a 2A rating.
- Extinguishers and water drums exposed to freezing conditions must be protected from freezing.
- Do not remove or tamper with fire extinguishers installed on equipment or vehicles, or in other locations, unless authorized to do so or in case of fire. If you use a fire extinguisher, be sure it is recharged or replaced with another fully charged extinguisher.

Fire Prevention

In order to prevent fire, the following principles should be followed:

- Internal combustion engine-powered equipment must be located so that exhausts are away from combustible materials.

- Smoking is prohibited at, or in the vicinity of, operations that constitute a fire hazard. Such operations must be conspicuously labeled "no smoking or open flame."
- Portable battery-powered lighting equipment must be approved for the type of hazardous locations encountered.
- Combustible materials must be piled no higher than 20 feet. Depending on the stability of the material being piled, this height may be reduced.
- Keep driveways between and around combustible storage piles at least 15 feet wide and free from accumulation of rubbish, equipment, or other materials.
- Portable fire-extinguishing equipment, suitable for anticipated fire hazards on the jobsite, must be provided at convenient, conspicuously accessible locations.
- Firefighting equipment must be kept free from obstacles, equipment, materials, and debris that could delay emergency use of such equipment. Become familiar with the location and use of the project's firefighting equipment.
- Discard and/or store all oily rags, waste, and similar combustible materials in metal containers on a daily basis.
- Storage of flammable substances in equipment or vehicles is prohibited unless such unit has an adequate storage area designed for such use.

Managing Fire Safety

The owner, manager, or employer of workers in a building or facility should have the answers to these questions:

- Do you have a fire emergency plan?
- Has it been reviewed and approved by the fire department?
- Have tenants and employees been given instruction on the details of the plan?
- Can the facility be evacuated to the street without interfering with fire department personnel?
- If the previous answer is no, are there areas of refuge at the facility?
- Are there provisions for physically challenged people who may be present?
- If a fire starts, will it be detected promptly? How?
- Will the fire department be notified promptly? How?
- Is there a provision for heating, ventilation, and air-conditioning smoke control?
- Is there an emergency communications system?
- Does the facility have area floor wardens? Have they been trained?
- Are firefighting equipment, emergency generators, and lighting systems ready to use if needed?
- Are all existing doors and exits clear?
- Are emergency hoses and fire extinguishers in working order?
- Will security measures, such as locking of doors, interfere with evacuation of occupants or access of firefighters?

- Is the fire department familiar with the facility in all aspects that would be helpful during an emergency?
- Has space been designated for a fire department command center at the facility?

The Importance of Fire Safety and Life Safety Design

The Life Safety Code (NFPA 101) from the National Fire Protection Association provides minimum requirements for the design, operation, and maintenance of a building or facility for safety to life from fire and similar emergencies. The code requires new and existing buildings or facilities to allow for "prompt escape" or to provide people with a reasonable degree of safety through other means.

The Life Safety Code meets its objective by following two parallel approaches. First, it defines hazards, along with general requirements for the means of egress (a path of exit travel to a public way outside), fire protection features (such as fire doors), and building or facility service and fire protection equipment (heating, ventilating, and air-conditioning systems; sprinkler systems; or fire detection systems, for example). Next, the Life Safety Code sets out life safety requirements, which vary with a building's or facility's use.

Unique among fire safety codes, the Life Safety Code has different provisions, depending on the type of occupancy and whether the building or facility is a new or existing construction. The Life Safety Code can be used in conjunction with a building code or alone in jurisdictions that do not have building or facility codes in place.

The Life Safety Code's objective is to provide safety to life during emergencies. However, two additional very positive spin-offs grow out of this objective.

- Many requirements that are designed to protect people also protect property, reducing the dollar loss associated with fire.
- Requirements that are designed to provide prompt escape during emergencies make buildings or facilities more pleasant during normal conditions. Spacious corridors and the convenience of multiple exits, for example, result from the requirement for prompt escape.

The Life Safety Code requires unlocked and unobstructed exits, multiple exits, fire doors, and regular fire exit drills. These key provisions state the following:

- Locks and hardware on doors shall be installed to permit free escape.
- Exits must be marked with a sign that is readily visible.
- Any door that is a means of egress must be capable of swinging from any direction to the full use of the opening. The door must swing in the direction of egress when serving a room or area with 50 or more occupants.
- Evacuation signals must be audible and visible.
- Fire drills should be conducted on a regular basis. All stairway exits should be accessible and marked as such since elevators may not function during a fire, or may expose passengers to heat, gas, and smoke. All stairwells should be separated/sealed from the main occupancy areas and have their own ventilation system.

Also, the following procedure should be followed:

- Heat-producing equipment—copiers, hot plates, and coffee makers—are often overlooked as potential fire hazards. Keep them away from anything that might burn.

- Electrical appliances can be fire hazards. Be sure to turn off all appliances at the end of the day. Use only grounded appliances plugged into grounded outlets (three-prong plug).

- If electrical equipment malfunctions or gives off a strange odor, disconnect it and call the appropriate maintenance personnel. Promptly disconnect and replace cracked, frayed, or broken electrical cords.

- Keep extension cords clear of doorways and other areas where they can be stepped on or chafed, and never plug one extension cord into another. Never place rugs or other materials on extension cords since they will not be able to dissipate the heat from the flow of electrical current.

- Smoke only where permitted, and have large nontip ashtrays and empty them only when ashes, matches, and smoking materials are cold. Make sure that no one, including visitors, has left cigarettes smoldering in wastebaskets or on furniture.

- Keep storage areas, stairway landings, and other out-of-the-way locations free of wastepaper, empty boxes, dirty rags, and other material that could fuel a fire or hamper an escape.

- Arson is the largest single cause of fires in buildings and facilities. Therefore, proper security measures to keep unauthorized individuals out of the building or facility will help prevent both theft and fire. In addition, make sure that alleys and other areas around buildings or facilities are well lit.

Through a program of scheduled inspections, unsafe conditions can be recognized and corrected before they lead to fires. A few moments each day should be taken to walk through the work areas. Look for items that have the potential to be fire hazards.

FPP Requirements

An FPP must be in writing, be kept in the workplace, and be made available to employees for review. However, an employer with 10 or fewer employees may communicate the plan orally to employees [29 CFR 1910.39(b)]. At a minimum, the FPP must include the following:

- A list of all major fire hazards, proper handling and storage procedures for hazardous materials, potential ignition sources and their control, and the type of fire protection equipment necessary to control each major hazard [29 CFR 1910.38(b)(2)(i)].

- Procedures to control accumulations of flammable and combustible waste materials [29 CFR 1910.38(b)(3)].

- Procedures for regular maintenance of safeguards installed on heat-producing equipment to prevent the accidental ignition of combustible materials [29 CFR 1910.38(c)(5)].

- The name and job title of employees responsible for maintaining equipment to prevent or control sources of ignition or fires [29 CFR 1910.38(b)(2)(ii)].
- The name or job title of employees responsible for the control of fuel source hazards [29 CFR 1910.38(b)(2)(ii)].
- An employer must inform employees, upon initial assignment to a job, of the fire hazards to which they are exposed. An employer must review with each employee the fire prevention necessary for self-protection [29 CFR 1910.38(b)(4)(i–ii)].

The elements of an FPP are as follows:

- Identification of the major workplace fire hazards and their proper handling and storage
- Potential ignition sources (e.g., smoking) and their control procedures, and the type of fire protection equipment or systems that can be used to control a fire
- Names or regular job titles of personnel responsible for fire suppression equipment or systems
- Names or regular job titles of personnel responsible for controlling fuel source hazards

Fire Protection Summary

Every year, there are about 7,000 fires that break out in high-rise office buildings, causing deaths, injuries, and millions of dollars in fire damage. Most of these could be eliminated if everyone practiced good fire prevention on the job and planned ahead for a fire emergency.

Although workers can suffer burns from a fire, most injuries and deaths are the result of smoke inhalation, breathing toxic materials, and the presence of carbon monoxide.

The causes of fires in an industrial environment are often the following:

- Electrical causes (22%)—lax maintenance in wiring, motors, switches, lamps, and heating elements
- Matches and smoking (18%)—near flammable liquids, stored combustibles, etc.
- Friction (11%)—hot bearings, misaligned or broken machine parts, choking or jamming materials, poor adjustment of moving parts
- Hot surfaces (9%)—exposure of combustibles to furnaces, hot ducts or flues, electric lamps or heating elements, and hot metal
- Overheated materials (7%)—abnormal process temperatures, materials in dryers, overheating of flammable liquids
- Open flames (6%)—gasoline or other torches, gas or oil burners
- Foreign substances (5%)—foreign material in stock
- Spontaneous heating (4%)—deposits in ducts and flues, low-grade storage, scrap waste, oily waste and rubbish
- Cutting and welding (4%)—highly dangerous in areas where sparks can ignite combustibles

- Combustion sparks (4%)—burning rubbish; foundry cupolas, furnaces, and fireboxes
- Miscellaneous (10%)—including incendiary cases, fires spreading from adjoining buildings, molten metal or glass, static electricity near flammable liquids, chemical action, and lighting

Spotting fire hazards in the work area is a matter of being familiar with the causes listed previously. Fire inspections should be conducted on a daily, weekly, monthly, etc., basis.

When a fire hazard is spotted, eliminate it immediately, if you are capable of doing so and have the authority to do so. Fill out a fire hazard report form and bring it to the supervisor's attention. If a fire has started, notify the appropriate personnel (company fire brigade, a supervisor, safety director, etc.) to give or sound a general alarm.

General guidance for fires and related emergencies is to immediately follow these procedures if a fire is discovered or you see/smell smoke:

1. Notify the local fire department.
2. Notify physical security or the building security force.
3. Be aware of fire suppression equipment or outlets.
4. Activate the building alarm (fire pull station). If not available or operational, verbally notify people in the workplace.
5. Isolate the area by closing windows and doors, and evacuate the facility, if it can be done so safely.
6. Shut down equipment in the immediate area, if possible.
7. If possible and if the worker has received appropriate training, use a portable fire extinguisher on the fire.
8. Assist oneself to evacuate.
9. Assist another to evacuate.
10. Control a small fire.
11. Do not collect personal or official items; leave the area of the fire immediately, and walk, do not run, to the exit and designated gathering area.
12. Provide the fire/police teams with the details of the problem upon their arrival. Special hazard information is essential for the safety of the emergency responders.
13. Do not reenter the building or facility until directed to do so. Follow any special procedures established for the unit's employees.
14. If the fire alarms are ringing in the building or facility, evacuate the building or facility and stay out until notified to return. Move to the designated meeting location or upwind from the building or facility, staying clear of streets, driveways, sidewalks, and other access ways to the building or facility. If you are a supervisor, try to account for employees, keep them together, and report any missing persons to the emergency personnel at the scene.

One result of the recent trend toward open work environments is that smoke from fires is not contained or isolated as effectively as in less open designs. Open designs allow smoke to spread quickly, and the incorporation of many synthetic and other combustible materials in facility fixtures (such as furniture, rugs, drapes, plastic wastebaskets, and vinyl-covered walls) often makes "smoky" fires. In addition to being smoky, many synthetic materials can emit toxic materials during a fire. For example, cyanide can be emitted from urethane, which is commonly used in upholstery stuffing. Most burning materials can emit carbon monoxide. Inhalation of these toxic materials can severely hamper a worker's chances of getting out of a fire in time. This makes it imperative for workers to recognize the signal to evacuate their work area and know how to exit in an expedient manner.

If an individual is overexposed to smoke or chemical vapors, remove the person to an uncontaminated area and treat for shock. Do not enter the area if a suspected life-threatening condition still exists (such as heavy smoke or toxic gases). If cardiopulmonary resuscitation (CPR) certified, follow standard CPR protocols. Get medical attention promptly.

If a person's clothing catches fire, extinguish the burning clothing by using the drop-and-roll technique, wrap the victim in a fire blanket, or douse the victim with cold water (use an emergency shower if it is immediately available). Carefully remove contaminated clothing; however, avoid further damage to the burned area. Cover the injured person to prevent shock. Get medical attention promptly.

Poor housekeeping contributes to an increased frequency of loss and greater loss potential. The added distribution of fuel does the following:

- Increases the probability of fire and explosion
- Causes a greater continuity of combustibles, making it easier for fire to spread
- Increases combustible loading by providing more fuel to feed a fire
- Creates the potential for dust explosions when dust accumulates
- Increases the probability of spontaneous ignition

In case of a fire, workers should be told to do the following:

- If a fire does break out, sound the alarm and call the fire department. Large fires start as small fires.
- Learn the sound of the building's fire alarm. Encourage management to schedule regular fire drills so that everyone will know how the alarm sounds and how to escape.
- Evacuation plans for the building should be posted where everyone can see them. These should be discussed with new employees during orientation.
- Learn the evacuation plan and participate in fire drills.
- Know the location of the two exits closest to their work area. Count the number of doors between their office and each of those exists—in case they must escape through a darkened, smoke-filled corridor where they can't see very well.
- Close the door to the room containing the fire and close all other doors that are passed through during their escape, assuming that you are the last person out. Closing the doors helps control the spread of fire.
- If it becomes necessary to use an escape route where there is smoke, crawl low under the smoke. Stay close to the floor where visibility is better, the air is less toxic, and it is cooler.
- Before opening a closed door, feel it with the back of the hand. If it is hot, don't open it. Use an alternate escape route. If it feels normal, open it carefully.

- Be ready to slam a door shut if heat or smoke starts to rush in. Once they are outside the building, move well away from the building to a designated meeting area where all members of the floor can be accounted for. If anyone is missing, notify the fire department. *Do not* reenter the building.

- If it's not possible to escape from the floor you are on, don't panic. Stay calm. Try to go to a room with an outside window and stay there. Try to keep smoke out and be sure doors are closed. Stuff the cracks around the door and vents using clothing, towels, paper, or whatever is available. If water is available, dampen a cloth and breathe through it to filter out smoke and gases. If there is a working telephone or cell phone, call the fire department and tell them exactly where you are. This information will be relayed immediately to the firefighters on the scene. Stay where you are and wave something to attract their attention.

- Each person with a disability should be assigned a coworker (and an alternate) to render assistance in case of an emergency. Participating in drills is especially important for those with disabilities.

- Never use an elevator during a fire emergency. Most modern elevators' select buttons are heat activated, so they might go to a floor where there is a fire and stop with doors wide open, exposing passengers to deadly heat and fumes.

- Be sure that stairwell doors are never locked.

A fire emergency can be devastating to a business. Even though most businesses are insured against fire losses, the loss of personnel, physical property, production, and customers can lead to the complete shutdown of the business. The end result is the impact on the community and vendors, subcontractor loss of business, and the fire's economic effect on these entities.

Since fire can transpire in the blink of an eye, steps must be taken to address this always-present risk. This requires a concerted effort to preplan for a fire event.

Further Readings

Reese, C.D. *Office Building Safety and Health*. Boca Raton, FL: CRC Press, 2004.

Reese, C.D. *Handbook of Safety and Health for the Service Industry, Volume 1: Industrial Safety and Health for Goods and Material Services*. Boca Raton, FL: CRC Press, 2009.

Reese, C.D. *Handbook of Safety and Health for the Service Industry, Volume 3: Industrial Safety and Health for Administrative Services*. Boca Raton, FL: CRC Press, 2009.

Reese, C.D. *Handbook of Safety and Health for the Service Industry, Volume 4: Industrial Safety and Health for People-Oriented Services*. Boca Raton, FL: CRC Press, 2009.

Reese, C.D. *Accident/Incident Prevention Techniques (Second Edition)*. Boca Raton, FL: CRC Press, 2012.

56

Hazardous Materials

Hazardous materials (abbreviated as HAZMAT or hazmat) have a total life cycle that begins at its birth and continues till an effective grave or permanent solution is found for disposal of it. Any time a hazmat is moved or transported, the risk of exposure is increased. It may manifest its danger during industrial mishaps, fires, explosions, or spills.

Hazmats, hazardous goods, or hazardous wastes (HWs) are solids, liquids, or gases that can harm people, other living organisms, property, or the environment. They are usually subject to chemical regulations. In the United States, United Kingdom, and Canada, dangerous goods are more commonly known as hazmats. Hazmat teams are personnel specially trained to handle dangerous goods, which include materials that are radioactive, flammable, explosive, corrosive, oxidizing, asphyxiating, biohazardous, toxic, pathogenic, or allergenic. Also included are physical conditions such as compressed gases and liquids or hot materials, including all goods that contain such materials or chemicals or may have other characteristics that render them hazardous in specific circumstances.

In the United States, dangerous goods are often indicated by diamond-shaped signage on the item, its container, or the building where it is stored. The color of each diamond indicates its hazard; e.g., flammable is indicated with red, because fire and heat are generally of red color, and explosive is indicated with orange, because mixing red (flammable) with yellow (oxidizing agent) creates orange.

Laws and regulations on the use and handling of hazmats may differ depending on the activity and status of the material. For example, one set of requirements may apply to its use in the workplace, while a different set of requirements may apply to spill response, sale for consumer use, or transportation. Most countries regulate some aspect of hazmats.

The most widely applied regulatory scheme is that for the transportation of dangerous goods.

The Globally Harmonized System of Classification and Labeling of Chemicals (GHS) is an internationally agreed-upon system set to replace the various classification and labeling standards used in different countries. GHS will use consistent criteria for classification and labeling on a global level.

Dangerous goods are divided into nine classes (in addition to several subcategories) on the basis of the specific chemical characteristics producing the risk. These nine classes are as follows:

- Class 1—explosives
- Class 2—gases
- Class 3—flammable liquids
- Class 4—flammable solids
- Class 5—oxidizing agents and organic peroxides
- Class 6—toxic and infectious substances
- Class 7—radioactive substances

- Class 8—corrosive substances
- Class 9—miscellaneous (asbestos, dry ice, etc.)

Why Hazmats Need Special Handling

There are thousands upon thousands of chemicals, which each have their own physical and chemical properties. These are the reasons they require specific handling:

- Workers must be protected during all cycles when these chemicals are manufactured, used, handled, stored, and moved, or during disposal.
- Any excess chemicals, waste, and by-products must be handled, stored, and transported in a safe manner.
- Only those with knowledge and experience relevant to hazmats should be allowed to manage the hazards presented by them.
- It takes specially trained individuals to handle hazmats, who are qualified to wear the appropriate protective equipment.
- Information must be available for each chemical regarding its use and handling.
- These chemicals are so pervasive throughout society that special precautions must be taken to protect the general public during manufacturing, use, storage, transport, and disposal.
- Some of these chemicals are so persistent once in the environment that they will always exist and can only be contained at best.
- Some chemicals can have immediate impacts, while others can be an issue in the future.

Transportation of Hazmats

Transporting hazmats increases exposure and potential risk and dangers. Hazmats in transportation must be placarded and have specified packaging and labeling. Some materials must always be placarded, while others may only require placarding in certain circumstances.

Trailers of goods in transport are usually marked with a four-digit United Nations Location Code (UN) number. This number, along with standardized logs of hazmat information, can be referenced by first responders (firefighters, police officers, and ambulance personnel), who can find information about the material in the *Emergency Response Guidebook*.

Different standards usually apply for handling and marking hazmats at fixed facilities, including diamond markings (a consensus standard often adopted by local governmental jurisdictions), Occupational Safety and Health Administration (OSHA) regulations requiring chemical safety information for employees, and Consumer Product Safety Commission (CPSC) requirements requiring informative labeling for the public, as well as wearing hazmat suits when handling hazmats.

Storage and Packing

It is a requirement that hazmats be packed and stored in appropriate designed containers, which are to prevent/contain spills during transportation mishaps. These containers should have the proper labels attached to them to allow for readly identification.

Transporting

One of the transport regulations is that, as a form of assistance during emergency situations, written instructions on how to deal with hazmats need to be carried and easily accessible in the driver's cabin. A license or permit card for hazmat training must be presented when requested by officials.

Dangerous goods shipments also require a special declaration form prepared by the shipper. Among the information that is generally required includes the shipper's name and address; the consignee's name and address; descriptions of each of the dangerous goods, along with their quantity, classification, and packaging; and emergency contact information.

The mission of the Federal Motor Carrier Safety Administration (FMCSA) is to improve truck and bus safety on our nation's highways. That includes reducing the number of transportation incidents that involve hazmats and could potentially harm the public and the environment. Developing programs to accomplish these goals and increase the safety of hazmat transportation is the responsibility of the FMCSA Hazardous Materials (HM) Program. Transporting hazmats? Make sure you keep the load on the road and do your part to prevent cargo tank truck rollovers.

Regulations and Hazmats

Due to the increase in the threat of terrorism in the early twenty-first century after the September 11, 2001 attacks, funding for greater hazmat-handling capabilities was increased throughout the United States, in recognition that flammable, poisonous, explosive, or radioactive substances in particular could be used for terrorist attacks.

The Pipeline and Hazardous Materials Safety Administration regulates hazmat transportation within the territory of the United States by Title 49 of the Code of Federal Regulations (CFR).

OSHA regulates the handling of hazmats in the workplace as well as response to hazmat-related incidents, most notably through Hazardous Waste Operations and Emergency Response (HAZWOPER) regulations found in 29 CFR 1910.120.

In 1984, several agencies [OSHA, Environmental Protection Agency (EPA), US Coast Guard (USCG), and National Institute for Occupational Safety and Health (NIOSH)] jointly published the first *Hazardous Waste Operations and Emergency Response Guidance Manual*, which is available for download, or can be purchased from the US Government Printing Office, Pub. 85-115.

The EPA regulates hazmats as they may impact the community and environment, including specific regulations for environmental cleanup and for handling and disposal of waste hazmats. For instance, transportation of hazmats is regulated by the Hazardous Materials Transportation Act. The Resource Conservation and Recovery Act (RCRA) was also passed to further protect human and environmental health.

The CPSC regulates hazmats that may be used in products sold for household and other consumer uses.

Hazmats are defined and regulated in the United States primarily by laws and regulations administered by the EPA, OSHA, US Department of Transportation (DOT), and US Nuclear Regulatory Commission (NRC). Each has its own definition of a *hazmat*.

OSHA's definition includes any substance or chemical that is a "health hazard" or "physical hazard," including the following: chemicals that are carcinogens, toxic agents, irritants, corrosives, sensitizers; agents that act on the hematopoietic system; agents that damage the lungs, skin, eyes, or mucous membranes; chemicals that are combustible, explosive, flammable, oxidizers, pyrophorics, unstable reactive, or water reactive; and chemicals that, in the course of normal handling, use, or storage, may produce or release dusts, gases, fumes, vapors, mists, or smoke that may have any of the previously mentioned characteristics. (Full definitions can be found in 29 CFR 1910.1200.)

The EPA incorporates the OSHA definition and adds any item or chemical that can cause harm to people, plants, or animals when released by spilling, leaking, pumping, pouring, emitting, emptying, discharging, injecting, escaping, leaching, dumping, or disposing into the environment. (40 CFR 355 contains a list of over 350 hazardous and extremely hazardous substances.)

The DOT defines a hazmat as any item or chemical that, when being transported or moved in commerce, is a risk to public safety or the environment and is regulated as such under its Pipeline and Hazardous Materials Safety Administration regulations (49 CFR 100–199), which include the Hazardous Materials Regulations (49 CFR 171–180). In addition, hazmats in transport are regulated by the International Maritime Dangerous Goods Code, Dangerous Goods Regulations of the International Air Transport Association, Technical Instructions of the International Civil Aviation Organization, and US Air Force joint manual, *Preparing Hazardous Materials for Military Air Shipments.*

The NRC regulates materials that are considered hazardous because they produce ionizing radiation, which means those materials that produce alpha particles, beta particles, gamma rays, x-rays, neutrons, high-speed electrons, high-speed protons, and other particles capable of producing ions. This includes "special nuclear material," by-product material, and radioactive substances. (See 10 CFR 20.)

Hazardous Waste

HW is waste that poses substantial or potential threats to public health or the environment. It also is a waste product that no longer serves a useful purpose or must be disposed of in some fashion. In the United States, the treatment, storage, and disposal of HW are regulated under RCRA. HWs are defined under RCRA in 40 CFR 261, where they are divided into two major categories: characteristic wastes and listed wastes.

- Characteristic HWs are materials that are known or tested to exhibit one or more of the following four hazardous traits:
  - Ignitability
  - Reactivity

- Corrosivity
- Toxicity
- Listed HWs are materials specifically listed by regulatory authorities as HWs that are from nonspecific sources, specific sources, or discarded chemical products.

The requirements of RCRA apply to all the companies that generate HW as well as those companies that store or dispose of HW in the United States. Many types of businesses generate HW.

These wastes may be found in different physical states such as gaseous, liquid, or solid. An HW is a special type of waste because it cannot be disposed of by common means like other by-products of our everyday lives. Depending on the physical state of the waste, treatment and solidification processes might be required.

Worldwide, it has been estimated that more than 400 million tons of HWs are produced each year, mostly by industrialized countries.

US Hazardous Waste

A US facility that treats, stores, or disposes of HW must obtain a permit for doing so under RCRA. Generators and transporters of HW must meet specific requirements for handling, managing, and tracking waste. Through RCRA, Congress directed the EPA to create regulations to manage HW. Under this mandate, the EPA developed strict requirements for all aspects of HW management including the treatment, storage, and disposal of HW. In addition to these federal requirements, states may develop more stringent requirements that are broader in scope than the federal regulations. Furthermore, RCRA allows states to develop regulatory programs that are at least as stringent as RCRA, and after review by the EPA, the states may take over responsibility for the implementation of the requirements under RCRA. Most states take advantage of this authority, implementing their own HW programs that are at least as stringent, and in some cases are more stringent than the federal program.

Final Disposal of HW

Historically, some HWs were disposed of in regular landfills. This resulted in unfavorable amounts of hazmats seeping into the ground. These chemicals eventually entered natural hydrologic systems. Many landfills now require countermeasures against groundwater contamination. For example, a barrier has to be installed along the foundation of the landfill to contain the hazardous substances that may remain in the disposed waste. Currently, HWs must often be stabilized and solidified in order to enter a landfill and must undergo different treatments in order for them to be stabilized and disposed of. Most flammable materials can be recycled into industrial fuel. Some materials with hazardous constituents can be recycled, such as lead acid batteries.

Recycling

Many HWs can be recycled into new products. Examples may include lead acid batteries or electronic circuit boards. When heavy metals in these types of ashes go through the proper treatment, they could bind to other pollutants and convert them into easier-to-dispose-of solids, or they could be used as pavement filling. Such treatments reduce the level of threat of harmful chemicals, like fly and bottom ash, while also recycling the safe product.

Portland Cement

Another commonly used treatment is cement-based solidification and stabilization. Cement is used because it can treat a range of HWs by improving physical characteristics and decreasing the toxicity and transmission of contaminants. The cement produced is categorized into five different divisions, depending on its strength and components. This process of converting sludge into cement might include the addition of pH adjustment agents, phosphates, or sulfur reagents to reduce the settling or curing time, increase the compressive strength, or reduce the leachability of contaminants.

Incineration, Destruction, and Waste-to-Energy

HW may be "destroyed." For example, by incinerating them at a high temperature, flammable wastes can sometimes be burned as energy sources. For example, many cement kilns burn HWs like used oils or solvents. Today, incineration treatments not only reduce the amount of HW but also generate energy from the gases released in the process. It is known that this particular waste treatment releases toxic gases produced by the combustion of by-product or other materials that can affect the environment. However, current technology has developed more efficient incinerator units that control these emissions to a point where this treatment is considered a more beneficial option. There are different types of incinerators, which vary depending on the characteristics of the waste. Starved-air incineration is another method used to treat HWs. Just like in common incineration, burning occurs; however, controlling the amount of oxygen allowed proves to be significant in reducing the amount of harmful by-products produced. Starved-air incineration is an improvement over the traditional incinerators in terms of air pollution. Using this technology, it is possible to control the combustion rate of the waste and therefore reduce the air pollutants produced in the process.

HW Landfill (Sequestering, Isolation, Etc.)

An HW may be sequestered in an HW landfill or permanent disposal facility. In terms of hazardous waste, a landfill is defined as a disposal facility or part of a facility where hazardous waste is placed or on land and which is not a pile, a land treatment facility, a surface impoundment, an underground injection well, a salt dome formation, a salt bed formation, an underground mine, a cave, or a corrective action management unit (40 CFR 260.10).

Summary

A hazmat is any item or agent (biological, chemical, radiological, and/or physical) that has the potential to cause harm to humans, animals, or the environment, either by itself or through interaction with other factors. Hazmat professionals are responsible for and properly qualified to manage such materials. This includes managing and/or advising other managers on hazmats at any point in their life cycle, from process planning and development of new products; through manufacture, distribution, and use; and to disposal, cleanup, and remediation.

Further Readings

Carson, R.A. and C.J. Mumford. *Hazardous Chemicals Handbook (Second Edition)*. Oxford: Butterworth Heinemann, 2002.

Northwestern University, ORS Laboratory Safety, *Chemical Hazards*. Evanston, IL. Available at http://www.nwu.edu/research-safety/chem/ethid2.htm, 2007.

Reese, C.D. *Material Handling Systems: Designing for Safety and Health*. New York: Taylor & Francis, 2000.

United States Department of Energy. *OSH Technical Reference Manual*. Washington, DC: US Department of Energy, 1993.

United States Department of Health and Human Services. NIOSH/OSHA/USCG/EPA. *Occupational Safety and Health Guidance Manual for Hazardous Waste Site Activities*. USGPO. Washington, DC: US Department of Health and Human Services, October, 1985.

United States Department of Labor. Occupational Safety and Health Administration. *General Industry Standards (29 CFR 1910)*. Washington, DC: US Department of Labor, 2006.

United States Department of Labor. Occupational Safety and Health Administration. *OSHA Handbook for Small Businesses (OSHA 2209)*. Washington, DC: US Department of Labor, 2006.

57

Transportation

Transportation may not be a philosophy or principle by definition, but we demonstrate the need for it to be addressed. Why address it as a topic in this book? Without burdening anyone with a mountain of data, transportation is a critical element in the modern world since people, materials, goods, food, and needed items all must be transported. Without efficient/safe/dependable transportation, the world as it is today would not exist.

It is expected by those using transportation of any kind that their well-being is insured by the designers and providers of transportation. As is shown by the record, this is not always the case. Usually, when transportation incidents occur, there is a great deal of damage to life, environment, and property. Transportation incidents can result in catastrophic events, with large loss of life and carnage from an airplane crash, loss of the mode of transportation (i.e., ship sinking), or environmental disasters such as fire and explosion from a tanker train derailment.

Employer's Responsibility

Employers in the transportation industry have two specific responsibilities. If they transport passengers of any type, then they have a direct liability for those individuals. This liability extends to a perceived high degree of safety and health when individuals are making use of their conveyance. Whether traveling by commercial air, by commuter or commercial rail, over water on ship/boats/cruise ships, on buses, on subways, or with commercial people-moving devices, those using such transportation make an inherent assumption that they are safe and that the employer has taken steps to insure such. If these personal conveyances are regulated, as most are, this sense of safety and health is usually accurate. But, when disaster strikes due to faulty equipment, faulty infrastructure, operator error, and/or an accident/incident event, then destruction and carnage ensue. Much effort is made by owners and employers to mitigate and reduce this potential risk. This is accomplished by finances, purchasing safe equipment, and instituting safety systems and process.

Transporting Materials

Employers who transport any type of materials, whether hazardous or benign, are still responsible for providing safe equipment (trucks, etc.) and safe operators (pilots, etc.) that follow all safety and health rules, and follow the safety and health regulation for that type of transportation. As materials are moved from one location to another, the risk of a mishap increases. At times, materials and people are transported as one entity in a mode of transportation. In most cases, hazardous materials and people are not mixed in the same transport. It is sufficient to say that the material being transported determines much of the risk. Normally, it would be expected that the most newly manufactured goods and regular freight items would be less hazardous when being moved. Since moving any material increases exposure and the risk from it, precautions must be taken to avoid creating any issues for the public and workers during transport of any materials.

The types of materials that require transport in some fashion may range from oil to chemicals, hazardous materials, large/awkward/heavy materials, radioactive, coal, raw/bulk materials, timber, and toxic chemicals. If any of these present a hazard to the public, such material on railroads traveling across the country or through cities will require special handling and precautions.

Probably the most overall consistent exposure and risk is on the highway system. One major reason is that a mixing of commercial, public, and pleasure transport is occurring. This one fact is the reason that the highways are so hazardous to drive on. There is no way to separate well-trained drivers from those with far less skills. Not everyone on these roadways is playing or observing the same set of rules, even if they are assumed to be. There seem to be no standards for the type of vehicles, size of vehicles, speed of travel, and common driving practices. In fact, it often seems to be a free-for-all that equates to every man for himself/herself. Although this is not actually the case, with so many vehicles on our roads going in all different directions, it is riskier to be on the roadways than any other activity undertaken by our citizens and employees of our business.

Employees of Transportation Businesses

A second point of emphasis in the transportation industry is to assure that the employees are properly protected from the hazards they face. This is accomplished in a similar manner as with other industries. They should be trained in the specific hazards of their unique part of the transportation industry, as well as safety, maintenance, and heath issues. All operators, drivers, captains, and engineers should receive job-specific training expectations, and the employer must assure that all have the operator skills to perform their jobs.

When employees are required to repair infrastructure such as replace rails or conduct maintenance activities, all of these actions should be done in a safe and healthy manner. Also, employees who are responsible for material handling must be trained in safe loading and unloading techniques. Employers must maintain a safe and healthy workplace for their employees.

Summary

Nothing will happen, nothing can happen, and nothing is happening without a transportation system. Food will not flow from the farms, employees will not arrive at work, mail will not be delivered, no tourism will happen, service workers and other services will not be available, and parts, goods, and products will not arrive. Our world as we know it will shut down. Safe transportation is mandatory for our survival.

Further Readings

Daugherty, J.E. *Industrial Safety Management: A Practical Approach.* Rockville, MD: Government Institutes, 1999.

Goetsch, D.L. *Occupational Safety and Health for Technologists, Engineers, and Managers (Fifth Edition).* Upper Saddle River, NJ: Pearson Prentice Hall, 2005.

Kohn, J.P. and T.S. Ferry. *Safety and Health Management Planning.* Rockville, MD: Government Institutes, 1999.

Lack, R.W. *Safety, Health, and Asset Protection: Management Essentials (Second Edition).* Boca Raton, FL: Lewis Publishers, 2002.

National Safety Council, *Motor Fleet Safety Manual (Third Edition).* Itasca, IL: NSC, 1986.

Section M

Meshing Safety with Management Approaches

Over the years, many management approaches have come and gone. The burning question is, how does safety and health fit into each of these approaches? It always seems this component lags behind the implementation of any new management technique. But it is soon realized that occupational safety and health (OSH) must be an integral part of managing the business process.

The contents of this section are as follows:

Chapter 58—Process Safety Management

Chapter 59—Total Quality Management

Chapter 60—Lean Safety

58

Process Safety Management

Process safety management (PSM) is a systems approach that identifies high-risk potential hazards, most of which were initially chemically related. Now the definition has been broadened to encompass any process where the hazard cannot be designed out or controlled by safety devices supported by warning devices or by detailed procedures or training. There is a set of elements that are viewed as necessary to achieve PSM. These elements are as follows:

- Process safety information
- Process hazard/risk analysis
- Management of change
- Mechanical integrity
- Operating procedures
- Safe work practices
- Process safety reviews
- Training
- Contractors
- Emergency response
- Incident investigation
- Audits
- Employee participation
- Trade secrets

Process Development

Any safety and health process begins as an evolving entity. If safety and health is not part of the development of the process itself, the process will have inherent flaws. As engineers formulated their concept for a process, the design that takes place should incorporate occupational safety and health from the very beginning of a process. Each step of the process design, development, and implementation should be reviewed to assure that hazardous potentials have been analyzed to determine exposure prevalence, degrees of risk, energy sources, and how best to protect the workers.

Process Hazard Analysis

The process hazard analysis (PHA) is a thorough, orderly, systematic approach for identifying, evaluating, and controlling the hazards of processes involving highly hazardous chemicals. The employer must perform an initial PHA (hazard evaluation) on all processes covered by this standard. The PHA methodology selected must be appropriate to the complexity of the process and must identify, evaluate, and control the hazards involved in the process.

First, employers must determine and document the priority order for conducting PHAs based on a rationale that includes such considerations as the extent of the process hazards, the number of potentially affected employees, the age of the process, and the operating history of the process. All initial PHAs should be conducted as soon as possible, but at a minimum, the employer must complete no fewer than 25% by May 26, 1994; 50% by May 26, 1995; 75% by May 26, 1996; and all initial PHAs by May 26, 1997. Where there is only one process in a workplace, the analysis must be completed by May 26, 1994.

PHAs completed after May 26, 1987, that meet the requirements of the PSM standard are acceptable as initial PHAs. All PHAs must be updated and revalidated, based on their completion date, at least every 5 years.

The employer must use one or more of the following methods, as appropriate, to determine and evaluate the hazards of the process being analyzed:

- What-ifs
- Checklist
- What-ifs/checklist
- Hazard and operability study (HAZOP)
- Failure mode and effects analysis (FMEA)
- Fault tree analysis
- An appropriate equivalent methodology

A discussion of these methods of analysis is contained in OSHA 3133, *Process Safety Management—Guidelines for Compliance*. Whichever method(s) is used, the PHA must address the following:

- The hazards of the process.
- The identification of any previous incident that had a potential for catastrophic consequences in the workplace.
- Engineering and administrative controls applicable to the hazards and their interrelationships, such as appropriate application of detection methodologies to provide early warning of releases. Acceptable detection methods might include process monitoring and control instrumentation with alarms, and detection hardware such as hydrocarbon sensors.
- Consequences of failure of engineering and administrative controls.
- Facility siting.
- Human factors.
- A qualitative evaluation of a range of the possible safety and health effects on employees in the workplace if there is a failure of controls.

The Occupational Safety and Health Administration (OSHA) believes that the PHA is best performed by a team with expertise in engineering and process operations, and that the team should include at least one employee who has experience with and knowledge of the process being evaluated. Also, one member of the team must be knowledgeable in the specific analysis methods being used.

The employer must establish a system to address promptly the team's findings and recommendations; ensure that the recommendations are resolved in a timely manner and that the resolutions are documented; document what actions are to be taken; develop a written schedule of when these actions are to be completed; complete actions as soon as possible; and communicate the actions to operating, maintenance, and other employees whose work assignments are in the process and who may be affected by the recommendations or actions.

At least every 5 years after the completion of the initial PHA, the it must be updated and revalidated by a team meeting the standard's requirements to ensure that the hazard analysis is consistent with the current process.

Employers must keep on file and make available to OSHA, on request, PHAs and updates or revalidation for each process covered by PSM, as well as the documented resolution of recommendations, for the life of the process.

Operating Procedures

The employer must develop and implement written operating procedures, consistent with the process safety information, that provide clear instructions for safely conducting activities involved in each covered process. OSHA believes that tasks and procedures related to the covered process must be appropriate, clear, consistent, and most importantly, well communicated to employees. The procedures must address at least the following elements.

Steps for each operating phase:

- Initial start-up
- Normal operations
- Temporary operations
- Emergency shutdown, including the conditions under which emergency shutdown is required, and the assignment of shutdown responsibility to qualified operators to ensure that emergency shutdown is executed in a safe and timely manner
- Emergency operations
- Normal shutdown
- Start-up following a turnaround, or after an emergency shutdown

Operating limits:

- Consequences of deviation
- Steps required to correct or avoid deviation

Safety and health considerations:

- Properties of, and hazards presented by, the chemicals used in the process
- Precautions necessary to prevent exposure, including engineering controls, administrative controls, and personal protective equipment
- Control measures to be taken if physical contact or airborne exposure occurs
- Quality control for raw materials and control of hazardous chemical inventory levels
- Any special or unique hazards
- Safety systems (e.g., interlocks, detection or suppression systems) and their functions.

Safe Work Practices

To ensure that a ready and up-to-date reference is available, and to form a foundation for needed employee training, operating procedures must be readily accessible to employees who work in or maintain a process. The operating procedures must be reviewed as often as necessary to ensure that they reflect current operating practices, including changes in process chemicals, technology, equipment, and facilities. To guard against outdated or inaccurate operating procedures, the employer must certify annually that these operating procedures are current and accurate.

The employer must develop and implement safe work practices to provide for the control of hazards during work activities such as lockout/tagout; confined space entry; opening process equipment or piping; and control over the entrance into a facility by maintenance, contractor, laboratory, or other support personnel. These safe work practices must apply both to employees and to contractor employees.

Employee Participation

Employers must develop a written plan of action to implement the employee participation required by PSM. Under PSM, employers must consult with employees and their representatives on the conduct and development of PHAs and on the development of the other elements of process management, and they must provide to employees and their representatives access to PHAs and to all other information required to be developed by the standard.

Training

Initial Training

OSHA believes that the implementation of an effective training program is one of the most important steps that an employer can take to enhance employee safety. Accordingly,

PSM requires that each employee presently involved in operating a process or a newly assigned process be trained in an overview of the process and in its operating procedures. The training must include emphasis on the specific safety and health hazards of the process, emergency operations including shutdown, and other safe work practices that apply to the employee's job tasks. Those employees already involved in operating a process on the PSM effective date do not necessarily need to be given initial training. Instead, the employer may certify in writing that the employees have the required knowledge, skills, and abilities to safely carry out the duties and responsibilities specified in the operating procedures.

Refresher Training

Refresher training must be provided at least every 3 years, or more often if necessary, to each employee involved in operating a process, to ensure that the employee understands and adheres to the current operating procedures of the process. The employer, in consultation with the employees involved in operating the process, must determine the appropriate frequency of refresher training.

Training Documentation

The employer must determine whether each employee operating a process has received and understood the training required by PSM. A record must be kept containing the identity of the employee, the date of training, and how the employer verified that the employee understood the training.

Contractors

Application

Many categories of contract labor may be present at a jobsite; such workers may actually operate the facility or do only a particular aspect of a job because they have specialized knowledge or skill. Others work only for short periods when there is a need for increased staff quickly, such as in turnaround operations. PSM includes special provisions for contractors and their employees to emphasize the importance of everyone taking care that they do nothing to endanger those working nearby who may work for another employer.

PSM, therefore, applies to contractors performing maintenance or repair, turnaround, major renovation, or specialty work on or adjacent to a covered process. It does not apply, however, to contractors providing incidental services that do not influence process safety, such as janitorial, food and drink, laundry, delivery, or other supply services.

Employer Responsibilities

When selecting a contractor, the employer must obtain and evaluate information regarding the contract employer's safety performance and programs. The employer also must inform contract employers of the known potential fire, explosion, or toxic release hazards related to the contractor's work and the process; explain to contract employers the applicable provisions of the emergency action plan; develop and implement safe work practices to control the presence, entrance, and exit of contract employers and contract employees in covered process areas; evaluate periodically the performance of contract employers in fulfilling their obligations; and maintain a contract employee injury and illness log related to the contractor's work in the process areas.

Contract Employer Responsibilities

The contract employer must do the following:

- Ensure that contract employees are trained in the work practices necessary to perform their job safely
- Ensure that contract employees are instructed in the known potential fire, explosion, or toxic release hazards related to their job and the process, and in the applicable provisions of the emergency action plan
- Document that each contract employee has received and understood the training required by the standard by preparing a record that contains the identity of the contract employee, the date of training, and the means used to verify that the employee understood the training
- Ensure that each contract employee follows the safety rules of the facility including the safe work practices required in the operating procedures section of the standard
- Advise the employer of any unique hazards presented by the contract employer's work

Pre-Start-Up Safety Review

It is important that a safety review takes place before any highly hazardous chemical is introduced into a process. PSM, therefore, requires the employer to perform a pre-start-up safety review for new facilities and for modified facilities when the modification is significant enough to require a change in the process safety information. Prior to the introduction of a highly hazardous chemical to a process, the pre-start-up safety review must confirm the following:

- Construction and equipment are in accordance with design specifications.
- Safety, operating, maintenance, and emergency procedures are in place and are adequate.
- A PHA has been performed for new facilities, recommendations have been resolved or implemented before start-up, and modified facilities meet the management of change requirements.
- Training of each employee involved in operating a process has been completed.

Mechanical Integrity

OSHA believes that it is important to maintain the mechanical integrity of critical process equipment to ensure it is designed and installed correctly and operates properly. PSM mechanical integrity requirements apply to the following equipment:

- Pressure vessels and storage tanks
- Piping systems (including piping components such as valves)
- Relief and vent systems and devices
- Emergency shutdown systems
- Controls (including monitoring devices and sensors, alarms, and interlocks)
- Pumps

The employer must establish and implement written procedures to maintain the ongoing integrity of process equipment. Employees involved in maintaining the ongoing integrity of process equipment must be trained in an overview of that process and its hazards and trained in the procedures applicable to the employees' job tasks.

Inspection and testing must be performed on process equipment, using procedures that follow recognized and generally accepted good engineering practices. The frequency of inspections and tests of process equipment must conform to manufacturers' recommendations and good engineering practices, or exceed these if determined to be necessary by prior operating experience. Each inspection and test on process equipment must be documented, identifying the date of the inspection or test, the name of the person who performed the inspection or test, the serial number or other identifier of the equipment on which the inspection or test was performed, a description of the inspection or test performed, and the results of the inspection or test.

Equipment deficiencies outside the acceptable limits defined by the process safety information must be corrected before further use. In some cases, it may not be necessary that deficiencies be corrected before further use, as long as deficiencies are corrected in a safe and timely manner, when other necessary steps are taken to ensure safe operation.

In constructing new plants and equipment, the employer must ensure that equipment as it is fabricated is suitable for the process application for which it will be used. Appropriate checks and inspections must be performed to ensure that equipment is installed properly and is consistent with design specifications and the manufacturer's instructions.

The employer also must ensure that maintenance materials, spare parts, and equipment are suitable for the process application for which they will be used.

Hot Work Permit

A permit must be issued for hot work operations conducted on or near a covered process. The permit must document that the fire prevention and protection requirements in OSHA regulations [1910.252(a)] have been implemented prior to beginning the hot work operations; it must indicate the date(s) authorized for hot work and identify the object on which hot work is to be performed. The permit must be kept on file until completion of the hot work.

Management of Change

OSHA believes that contemplated changes to a process must be thoroughly evaluated to fully assess their impact on employee safety and health and to determine needed changes to operating procedures. To this end, the standard contains a section on procedures for managing changes to processes. Written procedures to manage changes (except for "replacements in kind") to process chemicals, technology, equipment, and procedures, and changes to facilities that affect a covered process, must be established and implemented. These written procedures must ensure that the following considerations are addressed prior to any change:

- The technical basis for the proposed change
- Impact of the change on employee safety and health
- Modifications to operating procedures
- Necessary time period for the change
- Authorization requirements for the proposed change

Employees who operate a process and maintenance and contract employees whose job tasks will be affected by a change in the process must be informed of, and trained in, the change prior to start-up of the process or start-up of the affected part of the process. If a change covered by these procedures results in a change in the required process safety information, such information also must be updated accordingly. If a change covered by these procedures changes the required operating procedures or practices, they also must be updated.

Incident Investigation

A crucial part of the PSM program is a thorough investigation of incidents to identify the chain of events and causes so that corrective measures can be developed and implemented. Accordingly, PSM requires the investigation of each incident that resulted in, or could reasonably have resulted in, a catastrophic release of a highly hazardous chemical in the workplace.

Such an incident investigation must be initiated as promptly as possible, but not later than 48 hours following the incident. The investigation must be by a team consisting of at least one person knowledgeable in the process involved, including a contract employee if the incident involved the work of a contractor, and other persons with appropriate knowledge and experience to investigate and analyze the incident thoroughly.

An investigation report must be prepared including at least the following:

- Date of incident
- Date investigation began
- Description of the incident
- Factors that contributed to the incident
- Recommendations resulting from the investigation

A system must be established to promptly address and resolve the incident report findings and recommendations. Resolutions and corrective actions must be documented and the report reviewed by all affected personnel whose job tasks are relevant to the incident findings (including contract employees when applicable). The employer must keep these incident investigation reports for 5 years.

Emergency Planning and Response

If, despite the best planning, an incident occurs, it is essential that emergency preplanning and training make employees aware of, and able to execute, proper actions. For this reason, an emergency action plan for the entire plant must be developed and implemented in accordance with the provisions of other OSHA rules [29 CFR 1910.38(a)]. In addition, the emergency action plan must include procedures for handling small releases of hazardous chemicals. Employers covered under PSM also may be subject to the OSHA hazardous waste and emergency response regulation [29 CFR 1910.120(a), (p), and (q)].

Compliance Audits

To be certain that PSM is effective, employers must certify that they have evaluated compliance with the provisions of PSM at least every 3 years. This will verify that the procedures and practices developed under the standard are adequate and are being followed. The compliance audit must be conducted by at least one person knowledgeable in the process, and a report of the findings of the audit must be developed and documented noting deficiencies that have been corrected. The two most recent compliance audit reports must be kept on file.

Trade Secrets

Employers must make available all information necessary to comply with PSM to those persons responsible for compiling the process safety information, those developing the

PHA, those responsible for developing the operating procedures, and those performing incident investigations, emergency planning and response, and compliance audits, without regard to the possible trade secret status of such information. Nothing in PSM, however, precludes the employer from requiring those persons to enter into confidentiality agreements not to disclose the information.

OSHA's Response

Because of a large number of serious incidents, OSHA felt that a definitive standard was warranted. Thus, the OSHA process safety standard was promulgated in 29 CFR 1910.119. Its purpose was to prevent catastrophic accidents caused by the major releases of highly hazardous chemicals. To comply with this standard, industries must have written operating procedures, mechanical integrity, and formal incident investigation procedures. Other key elements are as follows:

- Coverage—The process safety standard is usually associated with large chemical and petrochemical processing plants. It is broader than that and can include other processes such as mechanical and electrical processes in industries such as manufacturing, aerospace, nuclear, transportation, and distribution.
- Employee participation—The standard requires that employees be involved in all aspects of the PSM program.
- Process safety information—Employees must be given access to any information maintained under the standard including chemical, process, and equipment data. The company must maintain these process safety information files.
- PHAs—PHAs are to be carried out by the employer on all processes covered by the standard. A PHA is to identify potential issues so that immediate corrective preventive action can be undertaken.
- Standard operating procedures (SOPs)—The standard requires that SOPs be established and maintained for safety, and applied to handling, processing, transporting, and storing of hazardous materials.
- Requirements for contractors—Contractors must have a comprehensive safety and health program, must understand the materials (i.e., chemicals) they need to work with and around, including emergency plans and other important information, occasionally evaluate their overall safety and health performance for compliance and performance, and maintain OSHA injury and illnesses records for contractors.

Summary

The standard includes a list of highly hazardous chemicals, which includes toxic, flammable, highly reactive, and explosive substances. The standard also contains specified minimum elements that the OSHA standard requires employers to comply with, as follows:

1. Develop and maintain written safety information identifying workplace chemical and process hazards, equipment used in the processes, and technology used in the processes

2. Perform a workplace hazard assessment, including, as appropriate, identification of potential sources of accidental releases, identification of any previous release within the facility that had a potential for catastrophic consequences in the workplace, estimation of workplace effects of a range of releases, and estimation of the safety and health effects of such a range on employees

3. Consult with employees and their representatives on the development and conduct of hazard assessments and the development of chemical accident prevention plans, and provide access to these and other records required under the standard

4. Establish a system to respond to the workplace hazard assessment findings, which shall address prevention, mitigation, and emergency responses

5. Review periodically the workplace hazard assessment and response system

6. Develop and implement written operating procedures for the chemical processes, including procedures for each operating phase, operating limitations, and safety and health considerations

7. Provide written safety and operating information for employees and employee training in operating procedures, by emphasizing hazards and safe practices that must be developed and made available

8. Ensure that contractors and contract employees are provided with appropriate information and training

9. Train and educate employees and contractors in emergency response procedures in a manner as comprehensive and effective as that required by the regulation promulgated pursuant to Section 126(d) of the Superfund Amendments and Reauthorization Act

10. Establish a quality assurance program to ensure that initial process-related equipment, maintenance materials, and spare parts are fabricated and installed consistent with design specifications

11. Establish maintenance systems for critical process-related equipment, including written procedures, employee training, appropriate inspections, and testing of such equipment to ensure ongoing mechanical integrity

12. Conduct pre-start-up safety reviews of all newly installed or modified equipment

13. Establish and implement written procedures managing change to process chemicals, technology, equipment, and facilities

14. Investigate every incident that results in or could have resulted in a major accident in the workplace, with any findings to be reviewed by operating personnel and modifications made, if needed

Further Readings

Bahr, N.J. *System Safety Engineering and Risk Assessment: A Practical Approach.* New York: Taylor & Francis, 1997.

Burns, T.E. *Serious Incident Prevention (Second Edition)*. Boston, MA: Gulf Professional Publishing, 1946.

Daugherty, J.E. *Industrial Safety Management: A Practical Approach*. Rockville, MD: Government Institutes, 1999.

Goetsch, D.L. *Occupational Safety and Health for Technologists, Engineers, and Managers (Fifth Edition)*. Upper Saddle River, NJ: Pearson Prentice Hall, 2005.

Kohn, J.P. and T.S. Ferry. *Safety and Health Management Planning*. Rockville, MD: Government Institutes, 1999.

Lack, R.W. *Safety, Health, and Asset Protection: Management Essentials (Second Edition)*. Boca Raton, FL: Lewis Publishers, 2002.

59

Total Quality Management

Total quality management (TQM) is a management effort to instill and make permanent a culture in which an organization continuously improves its ability to deliver high-quality products and services to customers. There is not a widely agreed-upon single approach for TQM efforts, but they typically draw heavily on the previously developed tools and techniques of quality control. It reached its highest popularity in the early 1990s and is now being replaced by ISO 9000, lean manufacturing, and Six Sigma management approaches.

Deming's 14 Points on Quality Management, a core concept on implementing TQM, is a set of management practices to help companies increase their quality and productivity.

Deming's 14 Points are as follows:

1. Create constancy of purpose for improving products and services.
2. Adopt the new philosophy.
3. Cease dependence on inspection to achieve quality.
4. End the practice of awarding business on price alone; instead, minimize total cost by working with a single supplier.
5. Improve constantly and forever every process for planning, production, and service.
6. Institute training on the job.
7. Adopt and institute leadership.
8. Drive out fear.
9. Break down barriers between staff areas.
10. Eliminate slogans, exhortations, and targets for the workforce.
11. Eliminate numerical quotas for the workforce and numerical goals for management.
12. Remove barriers that rob people of pride of workmanship, and eliminate the annual rating or merit system.
13. Institute a vigorous program of education and self-improvement for everyone.
14. Put everybody in the company to work accomplishing the transformation.

These TQM concepts can be put into place by any organization to more effectively implement TQM. As a TQM philosophy, W. Edwards Deming's work is foundational to TQM and its successor, quality management systems.

Because of the need to address quality and production competition in the United States, a new approach to managing such an effort was needed. Thus entered TQM.

Development of TQM in the United States

In the spring of 1984, an arm of the United States Navy asked some of its civilian researchers to assess statistical process control and the work of several prominent quality consultants and to make recommendations as to how to apply their approaches to improve the navy's operational effectiveness. The recommendation was to adopt the teachings of W. Edwards Deming. From the navy, TQM spread throughout the US federal government.

The private sector followed suit, flocking to TQM principles not only as a means to recapture market share from the Japanese, but also to remain competitive when bidding for contracts from the federal government since "total quality" requires involving suppliers, not just employees, in process improvement efforts.

Features of TQM

There is no widespread agreement as to what TQM is and what actions it requires of organizations; however, a review of the original United States Navy effort gives a rough understanding of what is involved in TQM.

The key concepts in the TQM effort undertaken by the navy in the 1980s include the following:

- Quality is defined by customers' requirements.
- Top management has direct responsibility for quality improvement.
- Increased quality comes from systematic analysis and improvement of work processes.
- Quality improvement is a continuous effort and conducted throughout the organization.

The navy used the following tools and techniques:

- The plan–do–check–act (PDCA) cycle to drive issues to resolution
- Ad hoc cross-functional teams (similar to quality circles) responsible for addressing immediate process issues
- Standing cross-functional teams responsible for the improvement of processes over the long term
- Active management participation through steering committees
- Use of the Seven Basic Tools of Quality to analyze quality-related issues:
  - **Cause-and-effect diagram** (also called Ishikawa or fishbone chart): identifies many possible causes for an effect or problem and sorts ideas into useful categories

- **Check sheet:** a structured, prepared form for collecting and analyzing data; a generic tool that can be adapted for a wide variety of purposes
- **Control charts:** graphs used to study how a process changes over time
- **Histogram:** the most commonly used graph for showing frequency distributions, or how often each different value in a set of data occurs
- **Pareto chart:** shows on a bar graph which factors are more significant
- **Scatter diagram:** graphs pairs of numerical data, one variable on each axis, to look for a relationship
- **Stratification:** a technique that separates data gathered from a variety of sources so that patterns can be seen (some lists replace *stratification* with *flowchart* or *run chart*)

Definition

While there is no generally accepted definition of TQM, several notable organizations have attempted to define it. These include the following:

> TQM in the Department of Defense is a strategy for continuously improving performance at every level, and in all areas of responsibility. It combines fundamental management techniques, existing improvement efforts, and specialized technical tools under a disciplined structure focused on continuously improving all processes. Improved performance is directed at satisfying such broad goals as cost, quality, schedule, and mission need and suitability. Increasing user satisfaction is the overriding objective.

The key words in describing TQM are the following:

- Management process
- Continuous improvement
- Harness both human and material resources
- Management approach
- Participation of all who are involved
- Customer satisfaction
- Everyone benefits from TQM
- Improving processes, products, services, and the culture

Although there is not a set-in-stone process for developing the TQM process, it is evaluated by how each entity of the organization is functioning in achieving quality management. The facets evaluated are leadership; strategy; customers; measurement, analysis, and knowledge management; workforce; operations; and results.

Principles of TQM

There are eight principles that most of those applying TQM seem to agree are imperative to be included when addressing TQM. They are as follows:

1. **Be customer focused:** The customer ultimately determines the level of quality. No matter what an organization does to foster quality improvement—training employees, integrating quality into the design process, upgrading computers or software, or buying new measuring tools—the customer determines whether the efforts were worthwhile.

2. **Insure total employee involvement:** All employees participate in working toward common goals. Total employee commitment can only be obtained after fear has been driven from the workplace, when empowerment has occurred, and when management has provided the proper environment. High-performance work systems integrate continuous improvement efforts with normal business operations. Self-managed work teams are one form of empowerment.

3. **Process centered:** A fundamental part of TQM is a focus on process thinking. A process is a series of steps that takes inputs from suppliers (internal or external) and transforms them into outputs that are delivered to customers (again, either internal or external). The steps required to carry out the process are defined, and performance measures are continuously monitored in order to detect unexpected variation.

4. **Integrated system:** All employees must know the business mission and vision. An integrated business system may be modeled by Malcolm Baldridge National Quality Award (MBNQA) or ISO 9000. Although an organization may consist of many different functional specialties often organized into vertically structured departments, it is the horizontal processes interconnecting these functions that are the focus of TQM.

5. **Strategic and systematic approach:** A critical part of the management of quality is the strategic and systematic approach to achieving an organization's vision, mission, and goals. This process, called strategic planning or strategic management, includes the formulation of a strategic plan that integrates quality as a core component.

6. **Continual improvement:** Using analytical, quality tools and creative thinking to become more efficient and effective, a major thrust of TQM is continual process improvement. Continual improvement drives an organization to be both analytical and creative in finding ways to become more competitive and more effective at meeting stakeholder expectations.

7. **Fact-based decision making:** Decision making must be *only* on data, not personal or situational thinking. In order to know how well an organization is performing, data on performance measures are necessary. TQM requires that an organization continually collect and analyze data in order to improve decision-making accuracy, achieve consensus, and allow prediction based on past history.

8. **Communication**: During times of organizational change, as well as part of day-to-day operation, effective communications plays a large part in maintaining morale and in motivating employees at all levels. Communications involve strategies, method, and timeliness.

TQM Implementation Approaches

TQM can't have just one effective solution for planning and implementing TQM concepts in all situations. The following is a list of generic models for implementing TQM theory:

- Train top management on TQM principles.
- Assess the current culture, customer satisfaction, and quality management system.
- Top management determines the core values and principles and communicates them.
- Develop a TQM master plan based on steps 1, 2, and 3.
- Identify and prioritize customer needs, and determine products or services to meet those needs.
- Determine the critical processes that produce those products or services.
- Create process improvement teams.
- Managers support the efforts by planning, training, and providing resources to the team.
- Management integrates changes for improvement in daily process management. After improvements, standardization takes place.
- Evaluate progress against the plan and adjust as needed.
- Provide constant employee awareness and feedback. Establish an employee reward/recognition process.

Safety and Health Integrated into TQM

The goal of continuous improvement by using TQM is compatible with occupational safety and health (OSH), whose goal is to improve the protection of the workforce from inherent hazards in the workplace using tried-and-true management techniques. Instead of having two management systems, OSH is compatible with the stated goals and principles of TQM to form a total overall management system. The integration can increase the success of TQM by incorporating some important aspects of OSH. They are as follows:

- System design, development, and training
- Evaluating and improving the safety process
- Improving employee participation and delegation
- Total safety management and delegation of responsibility
- Adaptive hazard management
- Innovative safety management
- Reducing employee errors
- Risk management

All of these are compatible with the total process approach that is both TQM and OSH. This can easily be adapted as one system approach. Integrating these two management systems will increase productivity and save some time and cost while preventing duplication of effort. As demonstrated by the previous pages of this book, the principles and philosophies of OSH are very much the same as those espoused by TQM.

The Concept of Continuous Improvement by TQM

TQM is mainly concerned with continuous improvement in all work, from high-level strategic planning and decision making to detailed execution of work elements on the shop floor. It stems from the belief that mistakes can be avoided and defects can be prevented. It leads to continuously improving results, in all aspects of work, as a result of continuously improving capabilities, people, processes, technology, and machine capabilities.

Continuous improvement must deal not only with improving results but also, more importantly, with improving capabilities to produce better results in the future. The five major areas of focus for capability improvement are demand generation, supply generation, technology, operations, and people capability.

A central principle of TQM is that mistakes may be made by people, but most of them are caused, or at least permitted, by faulty systems and processes. This means that the root cause of such mistakes can be identified and eliminated, and repetition can be prevented by changing the process.

There are three major mechanisms of prevention:

1. Preventing mistakes (defects) from occurring (mistake-proofing or poka-yoke)
2. Where mistakes can't be absolutely prevented, detecting them early to prevent them being passed down the value-added chain (inspection at source or by the next operation)
3. Where mistakes recur, stopping production until the process can be corrected, to prevent the production of more defects

Steps in Managing the Transition

In summary, first assess preconditions and the current state of the organization to make sure that the need for change is clear and that TQM is an appropriate strategy. Leadership styles and organizational culture must be congruent with TQM. If they are not, this should be worked on, or TQM implementation should be avoided or delayed until favorable conditions exist.

Remember that this will be a difficult, comprehensive, and long-term process. Leaders will need to maintain their commitment, keep the process visible, provide necessary support, and hold people accountable for results. Use input from stakeholders (clients, referring agencies, funding sources, etc.) as much possible, and of course, maximize employee involvement in design of the system.

Always keep in mind that TQM should be purpose driven. Be clear on the organization's vision for the future and stay focused on it. TQM can be a powerful technique for unleashing employee creativity and potential, reducing bureaucracy and costs, and improving service to clients and the community.

Summary

TQM is a method by which management and employees can become involved in the continuous improvement of the production of goods and services. It is a combination of quality and management tools aimed at increasing business and reducing losses due to wasteful practices.

TQM is a management philosophy that seeks to integrate all organizational functions (marketing, finance, design, engineering, production, customer service, etc.) to focus on meeting customer needs and organizational objectives.

Some of the companies that have implemented TQM include Ford Motor Company, Phillips Semiconductor, SGL Carbon, Motorola, and Toyota Motor Company.

TQM views an organization as a collection of processes. It maintains that organizations must strive to continuously improve these processes by incorporating the knowledge and experiences of workers. The simple objective of TQM is, "Do the right things right the first time, every time." TQM is infinitely variable and adaptable. Although originally applied to manufacturing operations, and for a number of years only used in that area, TQM is now becoming recognized as a generic management tool, just as applicable in service and public sector organizations. There are a number of evolutionary strands, with different sectors creating their own versions from the common ancestor. TQM is the foundation for activities that include the following:

- Commitment by senior management and all employees
- Meeting customer requirements
- Reducing development cycle times
- Just-in-time/demand-flow manufacturing
- Improvement teams
- Reducing product and service costs
- Systems to facilitate improvement
- Line management ownership
- Employee involvement and empowerment
- Recognition and celebration
- Challenging quantified goals and benchmarking
- Focus on processes/improvement plans
- Specific incorporation in strategic planning

This shows that TQM must be practiced in all activities, by all personnel and all types of businesses.

Further Readings

Gilbert, G. "Quality Improvement in a Defense *Organization*," *Public Productivity and Management Review*, 16(1), 65–75, 1992.

Hill, S. "Why Quality Circles Failed but Total Quality Management Might Succeed," *British Journal of Industrial Relations*, 29(4), 541–568, 1991.

Hyde, A. "The Proverbs of Total Quality Management: Recharting the Path to Quality Improvement in the Public Sector," *Public Productivity and Management Review*, 16(1), 25–37, 1992.

Ishikawa, K. *What Is Total Quality Control? The Japanese Way*. Englewood Cliffs, NJ: Prentice-Hall, 1985.

Martin, L. "Total Quality Management in the Public Sector," *National Productivity Review*, 10, 195–213, 1993.

Smith, A.K. "Total Quality Management in the Public Sector," *Quality Progress*, June 1993, 45–48, 1993.

Swiss, J. "Adapting TQM to Government," *Public Administration Review*, 52, 356–362, 1992.

Tichey, N. *Managing Strategic Change*. New York: John Wiley & Sons, 1983.

60

Lean Safety

In an effort to improve occupational safety and health (OSH), some have used a manufacturing process approach whose goal is continuous improvement. This is supposed to be accomplished by addressing cycle time from customer order to delivering a product by eliminating waste. Lean is a process review and improvement activity. This has taken a logical path that is working to incorporate safety with all other business activities. Lean and lean safety is the newest trend in business today. It is the word of the day, and it is lean.

New Approaches: Lean Safety and Sustainability

Lean safety is an evolving process that is gaining much support in the safety and health community. The lean process fosters engagement and ownership, where everyone is treated with dignity and contributes to successfully sustain it. With this in mind, safety and health can become a contributor and accepted component of continuous improvement, the core concept in lean.

Lean production strategy is utilized by companies and businesses to improve reliability, versatility, productivity, quality, and reduction of reworks and inventory. The reduction of waste through the entire production process results in finished products at the pace of customer demand with little or no waste. Safety must be integrated into the production process and recognized as a component of the waste reduction strategy since occupational injuries, illnesses, and deaths must be viewed as a waste of resources, which needs to be reduced.

Benefits of Lean Safety

As with any other new approach, there are the same advantages espoused by supporters such as lower internal resource requirements, lower cost, shorter start-up, better results in shorter time, broader distribution of information, less disruption to operations, greater support from workers and unions, greater internalization, and sustainability of techniques and results. This all sounds too good to be true, and it is bound to be more of a task than purported.

The manufacturer, consumer, or customer will experience a cleaner product, timely product, less waste, reduced risk of injury or illness (less personnel cost), conservation of resources both natural and man-made, more sustainability, and improvement all around, including safety. What more could you want?

The lean approach attempts to achieve its intentions by elimination/reduction of waste from overproduction, unnecessary transportation, waiting/queuing, extra processing, motion, inventory, and defects. All of this is meant to deliver the best possible value to the customer with the minimum resources possible.

This lean approach was developed by Toyota Production System (TPS) and has been around for a number of years. Linking lean and safety has become an issue for companies and individuals.

Challenges of Lean Safety

Lean is also fraught with issues for those trying to institute and implement. The philosophical concept of lean is confusing at best, while the vocabulary and rhetoric is like learning a new language, such as push/pull system, gemba, cycle time, kaizen blitz, A3, benchmarking, continuous flow, 5S, and PDCA. This new use of vocabulary has led to some misunderstandings.

Lean safety will require a change in company or business culture. When this transpires, you are not talking about months but years. People are always reticent to change. Lean is not sold as program but as a mindset. It requires involvement and ownership and a united effort to have everyone buy into it. This causes managers and supervisors to release control and become coaches and facilitators. Lean teams are to be organized, and each individual is viewed as a resource that adds value to the process.

Workers are often skeptical because lean conjures up visions of reduction of cost and people. Using safety has a balancing or leveling effect since safety is everyone's business. Lean safety requires total involvement in order to integrate it into the company's management process. Everyone can and should participate since they will be working with their own suggestions and recommendations.

Safety is often demonstrated by an organization's value in its employees. Companies that minimize risk and proactively improve their procedures stimulate a feeling of safety in the workplace. Safety should be a core value for all.

Lean safety will require training on lean itself and the tools used in the problem-solving techniques and goal of continuous improvement.

This leaner model is relatively easy to implement and provides greater and faster return than the resource-intensive, wait-and-see approaches. Every methodology that produces results must continue to evolve, or it will fail to produce new value. Whether you have an existing process or are exploring a new approach to safety improvement, there are many different methods to internalize this capability (i.e., public and private implementation and improvement workshops, hybrid implementations, fully supported implementations).

Changing Culture

Culture change starts at the top, where leadership will dwell on the process and change: not a blame game but how to fix what is a problem. Lean involves learning and empowering all employees. This will cultivate ownership and the development of support systems. Lean is engaging people and building trust.

Lean is a concept that is more intent upon changing culture with emphasis on involvement at all levels. It is a cerebral approach that emphasizes problem solving above minimal compliance. It is proactive, aimed at improving safety and reducing direct and indirect cost.

Again, it is a business process improvement process, not just a process. It is about developing lean thinkers and problem solvers. Most if all, it is about people and building trust. Trust is built by leadership behaviors and actions. Leaders are facilitators and coaches, and employees are the job experts, owners, and problem solvers. Leaders get trust by giving it.

The safety and health professional (S&HP) is charged with participating as a team member on teams to target the reduction of incidents, accidents, injuries, and property damage as components of waste in the process. The consequences of unsafe work environments are a source of waste and reduce efficiency of the production process. The S&HP is the team leader for safety and coordinates the safety and health effort of others by coaching and facilitating the safety component of the process. Culture change can be seen in how people think, act, and interact.

Learning and Training

The lean strategy for training and learning is that everyone needs to be involved because each employee brings with him/her life experiences, relevant knowledge, and practical approaches, which need to be channeled into the lean process. Lean will require the learning of a new vocabulary. Everyone will need to see how lean thinking is beneficial, how to use lean tools, how to accelerate improvement, how safety fits into lean activities, and how they fit in the lean process. This provides a baseline or knowledge base from which all can work.

By expanding employee knowledge and expertise and fostering lifelong learning, this will change the culture to problem solving and improving safety for everyone. Employees can make a difference. Employees who feel engaged and trusted make improvement and safety a daily priority.

Thus, the learning/training model provides the company with many new eyes to detect waste and begin the process of elimination. There is so much knowledge in numbers. The more active participants (employees), the more solutions and improvement possible.

Sustainability

There may be as many definitions of sustainability and sustainable development as there are groups trying to define it. All the definitions have to do with the following:

- Living within the limits
- Understanding the interconnections among economy, society, and environment
- Equitable distribution of resources and opportunities

However, different ways of defining sustainability are useful for different situations and different purposes. For this reason, various groups have created definitions of the following:

- Sustainability and sustainable development
- Sustainable community and society
- Sustainable business and production
- Sustainable agriculture

A simplistic definition seems to be maintenance of sustained production, safety, and waste reduction indefinitely.

Lean safety provides a process approach that is based upon ethical behavior and good business that helps create a culture that has the ability to mitigate safety crises such as chemical spills contaminating water, which could be brand killing or the demise of a business. It utilizes the principles of waste reduction while working to assure that its business process is both responsible and continuous in its intent to fit the social need and economic need to survive and exist over time.

Safety should fit into the company culture based on analytics and the commitment to continuous improvement, which is the basis of lean.

Planning is critical to a sustained approach to business. It entails a holistic approach to each facet of managing a business. Sustainable business culture takes a long-term view regarding humanity, environment, and safety as essentials that drive business success while assuring that resources are available to maintain a continuous improvement philosophy. With this said, it seems that lean and sustainability are members of the same family, as well they should be.

Sustainability is the capacity to endure and is dependent upon an integration of environmental factors, human factors, and economic factors. Such a concept involves environmental management and human demand for resources. Other factors include culture and political concerns as part of the matrix. Thus, as with lean, sustainability interfaces with economics through social and environmental consequences.

The first rule of sustainability is to align with natural forces or at least try not to defy them. Long-term profitability and efficient use of resources to sustain peoples' lives interjects the concept of safety into sustainability. Thus, lean and sustainability are bedfellows that cannot be easily separated.

Summary

Many companies view process improvement, safety, and waste reduction as separate programs. It would be more beneficial if lean, safety, and sustainability were seen as three strands of a single rope used to make the rope stronger and thus improve the process. Lean and lean safety can be useful in sustaining optimum production.

Lean strategies include even more vocabulary and concepts, such as just-in-time supply chain management, transformation from a push to a pull system, standardization, autonomation, and production leveling.

Many S&HPs are faced with incorporating OSH into the company's sustainable and lean approach to business. This requires the use and interpretation of a new vocabulary including new techniques in the integration process. To say the least, it is a confusing time for the S&HP since the fit is not well defined and is not a neat mesh into lean as was the old approach to safety and health. This will require the professional to look for the place where safety has a role. Thus, we now have lean safety as a part of this business approach.

Developing a culture of trust and ownership that engages others in learning, lean thinking, and problem solving for the greater good of all is an integral part of the continuous improvement of the process. Lean is a continuous process in itself and does not occur without commitment and a time-honored approach. The ultimate goal is sustainable, waste-free, safe, continuously improving production.

Further Readings

American Society for Quality. Profitable Applications of Value Stream Mapping Tutorial [Online], 2009. Accessed on February 22, 2013 at http://asq.org/learn-about-quality/lean/overview/value-stream-mapping.html

Aveta Business Institute. Explaining the Seven Types of Lean Waste [Online], 2012. Accessed on March 9, 2013 at http://www.sixsigmaonline.org/six-sigma-training-certification-information/articles/explaining-the-seven-types-of-lean-waste.html

Hafey, R.B. *Lean Safety: Transforming Your Safety Culture with Lean Management.* Boca Raton, FL: CRC Press, Taylor & Francis Group, 2010.

Hallowell, M., A. Veltri, and S. Johnson. *Safety & Lean: One Manufacturer's Lessons Learned and Best Practices.* Des Plaines, IL: Professional Safety, November, 2009.

Index

acceptable or tolerable risk, 197–199
  criteria, 198
  decision-making process, 198
  judgment calls, 198
  mathematical model, 198
  risk tolerated at acceptable level, 197–199
accident investigations, 51–55
  definition of accident, 51
  documentation of process, 54
  effectiveness, 52
  importance of, 51
  kinds of accidents, 52
  prompt reporting, 54
  recurrence of accidents, 52
  reporting accidents, 53
  reporting problems, 53
accident reconstruction, 189
airborne exposure, monitoring of, 121
American Match Co., 10
American Society of Safety Engineers (ASSE), 116
audits, *see* workplace inspections

background check (consultant), 127
behavior-based safety (BBS), 141–145
  barriers, 142–143
  BBS today, 141–142
  description, 142–143
  hindrances to implementing, 143–144
  program characteristics, 142
  purpose, 145
biological hazards, 24
biological stressors, 119
budget, 67–72
  compliance factors, 70–71
  controlling cost, 71–72
  environmental budgeting, 70
  health budgeting, 68
  identifiable categories, 68
  inaccuracy, 72
  items, 67–68
  management budgeting, 69
  need for budget, 67
  product safety budgeting, 70
  safety budgeting, 69
  written budget, 71
bullying, 157–164
  company policy, 163
  data regarding, 158–159

  facts about, 160–161
  plan establishment, 162
  prevention of, reason for, 162
  reasons for impacting work, 157
  why bullying occurs, 161–162
Bureau of Labor Statistics (BLS), 5, 17, 76

chemical exposure, dose of, 22
chemical hazards, 24
chemical stressors, 119
Clean Air Act Amendments of 1990, 62
commitment, *see* management commitment
  and involvement
committees, *see* labor/management (L/M)
  safety and health committees
common cause failure analysis, 189
communication, 151–155
  bulletin boards, 154
  communicator, 152–153
  computers, 154
  electronic signs, 154
  methods, 152
  motivation and, 136–137
  posters, 154–155
  public address system, 155
  by safety and health professional, 118
  safety talks, 155
  scenario, 151
  tools, 153–155
  training and, 61
  written materials, 153
consumer product safety screening, 318
controls, 213–219
  absorption, 214
  administrative controls, 216–217
  awareness devices, 216
  barriers, 214
  control at the level of the worker, 214
  control from hazard to worker, 214
  dilution, 214
  evaluating the effectiveness of controls,
    215–216
  hazard control summary, 218–219
  hazard prevention and controls, 218
  risk control, 215
  selecting controls, 215
  source control, 214
  technical aspect of hazard controls, 213

work practices, 216
work procedures, training, and supervision, 217
cost, 191–193
    cost of accidents, 192
    indirect costs, 192
    questions, 191
    true bottom line, 191
culture, *see* safety culture

dangerous goods, *see* hazardous materials (HAZMAT)
Deming's 14 Points on Quality Management, 361
design criteria analysis, 189
designing for prevention, 211–212
    designers' responsibility, 212
    designing for humans, 211

emergency planning, 275–281
    lack of plan, 280
    potential causes of emergencies, 276
    questions, 275
    reason for emergency action plans (EAPs), 277–278
    reason for preplanning, 280
    why certain elements are required or recommended, 278–279
employee involvement, 83–88
    caution flag, 83
    employer benefits, 83
    examples of, 84
    expectations, 85–86
    importance of employee outcomes, 86–87
    Pandora's box, 85
    reason for, 85–86
    why employees should be involved, 84
    why goals are needed, 87
    why management has to be involved, 84–85
    workplace issues, 86–87
energy
    classification of, 14
    kinetic, 14
    potential, 14
    sources of, 15
environmental budgeting, 70
environmental/occupational safety and health (EOSH), 70
ergonomics, 307–315
    assessing controls, 312–313
    awkward postures, 308
    cold temperatures, 309–310

contact stress, 309
developing an ergonomic program, 310–311
ergonomic controls, 311–312
extent of the problem, 310
force, 307
identifying controls, 312
implementing controls, 313
limits of exposure, 311
physical work activities and conditions, 311
proactive ergonomics, 314–315
repetition, 308
static postures, 308
tracking progress, 313–314
vibration, 308
ethics, 79–81
    accountability, 80
    compassion, 80
    credibility, 79
    fairness, 80
    honesty, 80
    integrity, 80
    as key principles to OSH, 79
    loyalty, 80
    occupational safety and health ethics, 81
    primary ethical values, 80
    promise keeping, 80
    respect, 80
    values, 79
external force, *see* terrorism

failure modes and effects analysis, 189
fire prevention and life safety, 323–333
    avoiding fires, 325–326
    causes of fires, 323–324
    fire prevention, 326–327
    fire prevention plan (FPP), 324
    fire protection summary, 330–333
    fire safety and protection, 326
    FPP requirements, 329–330
    importance of fire safety and life safety design, 328–329
    managing fire safety, 327–328
    reason to address fire hazards, 323
    what the OSHA standards require, 324–325
fleet safety, 261–265
    importance of driver/operator selection, 263
    purpose of program, 262
    reason for vehicle maintenance, 262–263
    records to be maintained, 264

Globally Harmonized System of Classification and Labeling of Chemicals (GHS), 335

hazard analysis, 177–181
    definition of, 177–178
    hazards and risk, 178
    job safety assessment, 180
    phase hazard analysis, 179–180
    preliminary hazard analysis, 179
    program, 179
    reasons for, 177–178
hazard controls, *see* controls
hazard identification, 173–175
    approach, 173
    benefits, 174
    management control of, 174
    for protection, 174
    responses, 173
    work site surveys, 174
hazardous materials (HAZMAT), 335–341
    dangerous goods, classes of, 335–336
    final disposal of HW, 339–340
    hazardous waste (HW), 335, 338–339
    HW landfill (sequestering, isolation, etc.), 340
    incineration, destruction, and waste-to-energy, 340
    Portland cement, 340
    recycling, 340
    regulations, 337–338
    special handling, 336
    transportation, 336–337
    US hazardous waste, 339
hazards, 13–16
    anticipating and predicting, 116
    biological, 24
    causes of accidents/injuries, 13
    chemical, 24
    definition, 13
    discovery of, 96
    energy, 14–15
    ergonomic, 24
    health, 22–25
    physical, 24
    safety, sources of, 18–19
    sources of energy, identification of, 13
    workplace stress, 24
health, 21–26
    biological hazards, 24
    budgeting, 68
    chemical exposure at work, 21–22
    chemical hazards, 24
    ergonomic hazards, 24

    false sense of security, 21
    hazard evaluations (HHEs), 229
    health hazard identification checklist, 25
    health hazard prevention, 23
    health hazards, 22
    identifying health hazards, 23–25
    investigative process, 25
    latency period, 21
    physical hazards, 24
    shift work, 24
    workplace stress, 24
Healthy People 2010 objectives, 4
Healthy People 2020, 23
hierarchy of needs (Maslow), 31
history, 9–11
    "common laws", 10
    evolution of OSH, 9–11
    first Acts and Regulations, 10
    Hammurabi, 9
    Hippocrates, 9
    Middle Ages, 9
    phossy jaw, 10
    results from history, 11
    workers' compensation laws, 10
hot work permit, 355–356
human factors, 303–305
    analysis, 189
    benefits, 304
    defining human factors, 303
    human characteristics, 304
    human factor safety, 303
    reason for human causal factors, 304

incentives/rewards, 167–170
    barrier example, 169
    fleeting incentives, 167
    incentive programs, 168–169
    incentives, 167–168
    peer pressure, 168
    types, 168
incident data, analysis of, 73–74
indirect cause (accidents/injuries), forms of, 18
industrial hygienist (IH), 119–122
    background, 120
    knowledge, 119
    monitoring of airborne exposure, 121
    need for, 120–121
    operations, 120
    reason for, 119–120
    workplace stressors, addressing of, 119
inspections, *see* workplace inspections
integrated accident event matrix, 189

job instruction training (JIT), 58
job safety analysis (JSA), 245–252
    change of frequency of performing a job,
        250
    change of job procedures, 249–250
    development of ways to eliminate or control
        hazards, 249
    development steps, 247
    human problems, 249
    identification of hazards associated with
        each job step, 248–249
    PPE use, 250–251
    questions, 246–247
    reason to develop, 245–246
    selection of jobs by using criteria,
        247–248
job safety observation (JSO), 253–260
    job or task selection for planned JSO,
        254–255
    need for checklist of activities to be
        observed, 256–257
    need for JSO preparation, 255
    paper form, 254
    postobservation, 258
    purpose, 253–254
    reason for observation, 257–258
    unsafe behaviors or poor performance,
        dealing with, 258–259
joint labor/management committees, *see* labor/
    management (L/M) safety and health
    committees

kinetic energy, 14

labor/management (L/M) safety and health
        committees, 89–94
    accomplishments, 93
    benefits, 92
    committee makeup, 90
    do's and don'ts, 90–91
    expectations, 91
    OSH committees, 92–93
    outcomes, 92
    procedures, 93–94
    purpose, 89
    record keeping, 90
lean safety, 369–373
    benefits, 369–370
    challenges, 370
    changing culture, 370–371
    learning and training, 371
    sustainability, 369, 371–372

life safety, *see* fire prevention and life safety
line supervisors, 107–108
    duties, 107–108
    evaluation form, 108
    as role model, 107

management, 29–35
    budgeting, 69
    decision to operate unsafely, 33
    failure to manage, 30
    functions, 29
    hierarchy of needs (Maslow), 31
    principles, 32
    questions, 33
    reason for managing OSH, 30, 34–35
    safety and health (managing), 30–31
    subsystems, 33
    why managing safety and health is a needed
        entity, 32–34
    workers' compensation, containment
        of premiums, 31
management commitment and involvement,
        103–106
    budget, 104
    entities held responsible, 105
    further readings, 106
    goals and objectives, 104
    roles and responsibilities, 105
    tough love, 104
    ways of conveying commitment, 106
materials and structural analysis, 189
Mine Safety and Health Administration
        (MSHA), 221
motivating safety and health, 131–139
    communications, 136–137
    defining motivation, 133
    essence of motivation, 139
    key person, 137
    motivational environment, 134
    principles of motivation, 133–134
    role model, 137
    setting the stage, 132
    simplify motivation, 136–137
    structuring the motivational environment,
        134–136
    traits critical to understanding, 138
    ways to motivate, 138

National Institute for Occupational
        Safety and Health (NIOSH), 221,
        228–229
National Safety Council (NSC), 5

National Traumatic Occupational Fatalities
    Surveillance System (NTOF), 5
nuclear, chemical, and biological (NBC) agents,
    293, 295

occupational health, *see* health
occupational safety, *see* safety (occupational)
occupational safety and health (OSH),
    introduction to, 3–8
    components of safety and health initiatives,
        5–7
    consequences of not addressing safety
        and health, 3
    data systems, 4–5
    hazard control factors, 6
    illness and injury investigations, 6
    list of considerations, 7
    management factors, 6
    motivational factors, 6
    need for occupational safety and health, 3–5
    philosophy, 4
    questions, 3–4
    record-keeping factors, 6
    workplace injuries and illnesses, 5
Occupational Safety and Health Act of 1970, 10,
    40
Occupational Safety and Health Administration
    (OSHA), 11, 221, 227–234
    citations and violations, costs for, 68
    database (citation costs), 70
    discrimination against workers, 231
    employer responsibilities under OSHAct,
        229–230
    fire prevention standards, 324–325
    importance of employee involvement, 83
    inspections, 232
    mandated training, 57
    National Institute for Occupational Safety
        and Health, 228–229
    occupational injuries and illnesses, 233
    Occupational Safety and Health Review
        Commission (OSHRC), 229
    PPE regulations, 222
    protections under the OSHAct, 228
    record keeping, 74
    regulation, compliance with, 95
    required written programs, 45–46
    right to information, 231–232
    standards, 228
    training, purpose of, 62
    workers' rights and responsibilities under
        OSHAct, 230–231
    worker training, 232–233

Occupational Safety and Health Review
    Commission (OSHRC), 229
off-the-job safety, 299–300
on-the-job training (OJT), 58

personal protective equipment (PPE), 221–225
    drawbacks, 221
    examples, 223
    OSHA regulations, 222
    program establishment, reason for, 223–224
    reason for hazard assessment, 223
phossy jaw, 10
physical hazards, 24
physical stressors, 119
potential energy, 14
preventive maintenance programs (PMPs),
    267–271
    components needed, 268
    formalized PMP, 269
    management's responsibility, 270
    need for maintenance, 270
    reason for management's role, 268–269
    reason for preventive maintenance, 268
    reason the operator should conduct
        inspections, 269
    reason to have PMP, 267
process safety management (PSM), 349–360
    compliance audits, 357
    contract employer responsibilities, 354
    contractors, 353–354
    emergency planning and response, 357
    employee participation, 352
    hot work permit, 355–356
    incident investigation, 356–357
    management of change, 356
    mechanical integrity, 355
    necessary elements, 349
    operating procedures, 351–352
    OSHA's response, 358
    pre-start-up safety review, 354
    process development, 349
    process hazard analysis (PHA), 350–351
    safe work practices, 352
    trade secrets, 357–358
    training, 352–353
product safety, 317–320
    budgeting, 70
    consultation, 318
    consumer product safety screening, 318
    development of safe design action plan, 318
    how safety screening is done, 319
    labeling, 319
    legal obligations, 318

principles of safe design, 317
reason a safe design approach should be
    employed, 317
why products are screened for safety, 319
programs, *see* safety and health programs;
    special emphasis programs

regulations
    HAZMAT, 337–338
    OSHA, required written programs, 45–46
    PPE, 222
rewards, *see* incentives/rewards
risk, *see* acceptable or tolerable risk
risk assessment, 201–203
    development, 201
    explanation, 202
    purpose, 202
    risk evaluation, 202
risk management, 205–207
    components, 205–206
    reason for, 206
root cause analysis, 183–185
    considerations, 183
    corrective action programs, 184
    functions, 184–185
    phases, 183–184
root cause analysis, forms of, 187–190
    basic root cause analysis, 188
    definition of, 187
    elimination of causes, 187
    methods, 188–189
    system safety engineering, 189
Russell Sage Foundation, 10

safe operating procedures (SOPs), 237–243
    benefits, 239
    components, 239–240
    definition, 237
    important function of, 242
    poorly written, 240–241
    problems and errors, 241
    uses, 237–238
    why SOPs work, 241–242
safety (occupational), 17–20
    definition of trauma, 17
    indirect cause, forms of, 18
    need for, 18
    principles, 19
    safety hazards, sources of, 18–19
    trauma events, 17
safety (toolbox) talks, 165–166
    advantages, 165
    documentation, 166

purposes, 165
speaker duties, 165–166
safety and health budget, *see* budget, safety
    budgeting
safety and health consultant, 123–128
    background check, 127
    hiring process, 126–127
    interviewing a consultant, 124–125
    need for, 123–124
    scope of work, 125–126
safety and health ethics, *see* ethics
safety and health professional, 115–118
    anticipating and predicting hazards, 116
    communication, 118
    duty, 117
    failure modes, 116
    need for, 115
    program development, 117
    responsibilities, 118
    titles, 115–116
safety and health programs, 37–47
    assessment tools, 43
    benefits, 45
    components, 40–41
    controversy, 37
    economic consideration, 39
    employer requirements, 38–39
    humanitarian consideration, 38
    legal obligation, 38
    purpose, 38–39
    questions, 37, 41–43
    reason for building, 39–40
    written programs, requirements for, 45–46
safety culture, 147–150
    assessment, 150
    characteristics, 150
    definition, 147
    developing or changing a safety culture,
        147–149
    factors impacting, 148
    motivator to change, 149
    positive safety culture, 149
safety director or manager, 113–114
    accountability, 114
    performance expectations, 113–114
    "safe production", 114
scope of work document (consultant),
    125–126
security, workplace, *see* workplace security
    and violence
shift work, 24
sneak circuit analysis, 189
software hazard analysis, 189

special emphasis programs, 49–50
  areas of focus, 49
  need for, 49
  purpose of, 49
statistics and tracking, 73–78
  ancillary data needed for more complete
    analysis, 75
  challenge in collecting data, 76
  company records, 74–75
  incident data, analysis of, 73–74
  OSHA record keeping, 74
  safety and health statistics data, 76
  statistical analysis for comparisons, 76
  tracking benefits, 73
  workers' compensation, 77
system safety engineering, 189

terrorism, 293–297
  hardening facilities increases protection,
    294–295
  potential terrorist's weapons, 295
  prevention, 296
  protection from chemical, biological,
    or radiological attacks, 295–296
  tools, 293
  travel security, 294
time loss analysis, 189
tolerable risk, *see* acceptable or tolerable risk
toolbox talks, *see* safety (toolbox) talks
total quality management (TQM), 361–368
  continuous improvement by, 366
  definition, 363
  Deming's 14 Points on Quality Management,
    361
  development of TQM in the United States,
    362
  features, 362–363
  implementation approaches, 365
  principles, 364
  safety and health integrated into,
    365–366
  steps in managing the transition, 366–367
tough love, 104
tracking, *see* statistics and tracking
trade secrets, 357–358
training, 57–64
  accident prevention, 58
  communication, 61
  failure to have trained workforce, 57
  importance of training, 61–62
  legal basis for training, 63
  OSHA training, purpose of, 62
  reasons to train employees, 60–61

reasons to train new employees, 58–59
reasons to train supervisors, 59–60
when to train, 57–58
transportation, 343–345
  employees of transportation businesses, 344
  employer's responsibility, 343
  transporting materials, 343–344
trauma
  definition of, 17
  events, observation of, 17
travel security, 294, 295
Triangle Shirtwaist Co. (1910), 10

US Department of Health and Human Services
    (DHHS), 4

values, *see* ethics
vehicle safety, *see* fleet safety
violence, workplace, *see* workplace security
    and violence

Walsh–Healey Act, 10
workers, 109–111
  discipline policy, 110
  employer expectations of, 109
  entities responsible for accident prevention,
    109
  reasons for worker responsibilities,
    109–110
workers' compensation
  basis of premium, 77
  containment of premiums, 31
  laws, 10
Work in America Institute study, 86
workplace
  chemical exposure in, 21–22
  culture, 148
  fatalities, statistics on, 11
  injuries and illnesses, toll of, 5
  issues, employee involvement and, 86–87
  stressors, 119
workplace inspections, 95–99
  hazard discovery, 96
  importance of, 95
  instrument types, 98
  need for, 95–97
  reasons for, 95
  what to inspect (list), 97–98
  when to inspect, 97
workplace security and violence, 283–291
  administrative controls, 286
  behavioral strategies, 286

cost of violence as a reason to address, 287
environmental design, 285–286
perpetrator and victim profile, 287
prevention efforts, 287–288
prevention strategies and security,
   285–286

reason for program development, 288–289
reason to address workplace violence, 290
risk factors, 285
types of workplace violence and events,
   289–290
workplace violence statistics, 283–284